Enzyme Kinetics Lecture Notes
Second Edition

Andreas Kukol

Copyright © 2015, 2017 Andreas Kukol

Create Space Independent Publishing, Charleston, USA

All rights reserved

ISBN-13: 978-1548471019
ISBN-10: 1548471011

Preface

The 'Lecture Notes' cover the topic of enzyme kinetics for a three-year undergraduate programme in bioscience. Many parts are relevant for all bioscience degree courses, such as pharmacology or biomedical sciences, while some of the advanced areas are more suitable for final year biochemistry, molecular biology or biotechnology courses.

The text does not assume much background knowledge except familiarity with the concepts of molar concentrations, chemical reactions and some basic mathematical concepts (the four basic arithmetic operations, fractions, exponents and logarithms, the notion of solving an equation). All other background maths and other concepts are explained in the text.

Various sections in the second edition have been expanded with textboxes providing additional explanations of spectrophotometry, calculations with the Arrhenius equation, how to derive rate equations for complex mechanisms, Cleland diagrams and matrices, eigenvalues and –vectors. Two new sections were added about enzyme inhibitor constants in pharmacology and allosteric enzymes.

Contents

1 Introduction 1

    1.1 Enzyme assays 2

    1.2 Application of enzyme assays 2

    1.3 Practical aspects 4

        1.3.1 Pre-steady state kinetic measurements 8

2 Chemical Kinetics 11

    2.1 Definition of the reaction rate 12

    2.2 Rate law and reaction mechanism 13

        2.2.1 Order of rate laws 13

        2.2.2 Relationship between rate laws and reaction mechanisms 14

    2.3 Integrating the rate law 16

    2.4 Rate laws for complex reaction mechanisms 20

    2.5 Compound rate laws 22

    2.6 The rate limiting step 22

    2.7 Temperature dependence of reaction rates 24

    2.8 A molecular picture of chemical reactions 27

3 Enzyme Kinetics 29

    3.1 The transition state theory applied to enzymes 29

    3.2 The reaction mechanism of alkaline phosphatase 30

    3.3 The Michalis-Menten equation 33

        3.3.1 Interpretation of the Michaelis-Menten equation 37

        3.3.2 Enzyme activity and enzyme units 39

    3.4 Analysis of enzyme kinetic data 40

        3.4.1 Obtain initial reaction rates 41

        3.4.2 The hyperbolic plot 42

        3.4.3 The Lineweaver-Burk plot 43

    3.4.4 Non-linear curve fitting 46

    3.4.5 Fitting of differential equations 47

3.5 Reversible inhibition 47

    3.5.1 Competitive inhibition 48

    3.5.2 Uncompetitive inhibition 49

    3.5.3 Mixed and non-competitive inhibition 51

    3.5.4 Data analysis for reversible inhibition 52

    3.5.5 Inhibitor binding sites for reversible inhibition 54

    3.5.6 Inhibitor constants in pharmacology 55

3.6 Deviations from Michaelis-Menten kinetics 58

    3.6.1 Substrate inhibition 58

    3.6.2 Product inhibition 59

    3.6.3 Sigmoid kinetics and cooperativity 60

3.7 Enzyme reactions with two substrates 62

    3.7.1 Ternary complex mechanisms 62

    3.7.2 Double displacement/ping-pong reaction 67

3.8 Allosteric enzymes 70

4 Single-Molecule Kinetics 75

4.1 Single molecule chemical kinetics 75

    4.1.1 Systems of unimolecular reactions 76

    4.1.2 Bimolecular reactions 82

4.2 Single molecule enzyme kinetics 83

    4.2.1 The single molecule Michaelis-Menten equation 85

4.3 Stochastic simulations 87

1 Introduction

Enzyme kinetics is concerned with the influence of enzymes on chemical reaction rates. Enzymes are *biological catalysts that speed up chemical reactions*. They occur naturally in biological systems, although in modern bio-technology they are also used in isolation to facilitate reactions of organic chemistry or to participate in the detection of analytes.

Kinetics of chemical reactions is a topic of physical chemistry concerned with chemical reaction rates, thus enzyme kinetics in particular belongs to the area of biophysical chemistry.

Enzymes are to a large extent protein molecules, although some enzymes are made from RNA and are referred to as ribozymes. An example of the three-dimensional structure of an enzyme is shown in Figure 1. Enzymes are nanometre sized particles that usually have a surface-accessible cleft or groove that forms the active site.

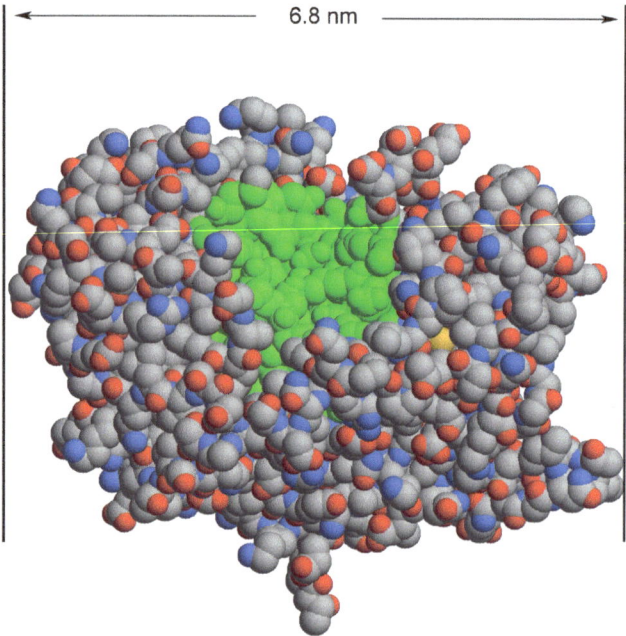

Figure 1: Three-dimensional structure of the enzyme alkaline phosphatase (PDB-ID: 1ALK). The active site is highlighted in green. The approximate diameter of this globular protein is 6.8 nm (nanometre).

Proteins, which form the majority of enzymes, are polymers of amino acids. Twenty different proteinogenous amino acids exist as the building blocks of proteins; amino acids are joined by the peptide bond into a long chain (449 amino acid residues in Figure 1) that is folded into a defined three-dimensional structure. The sequence of amino acids is determined by the gene encoding that particular protein. Often enzymes contain ions or organic molecules as *cofactors*. The alkaline phosphatase shown in Figure 1 contains two Zn^{2+} ions and one Mg^{2+} ion as cofactors.

In this introductory chapter some basic definitions, application examples and practical aspects of enzyme kinetics are given.

1.1 Enzyme assays

An enzyme assay is a laboratory method that measures enzyme activity, or in other words, that measures the rate of a chemical reaction catalysed by an enzyme. If an enzyme or a catalyst in general is involved in a chemical reaction, the reactant is normally called the *substrate* as illustrated schematically in Figure 2. Sometimes it may be the purpose of an enzyme assay to obtain information about enzyme inhibitors or activators.

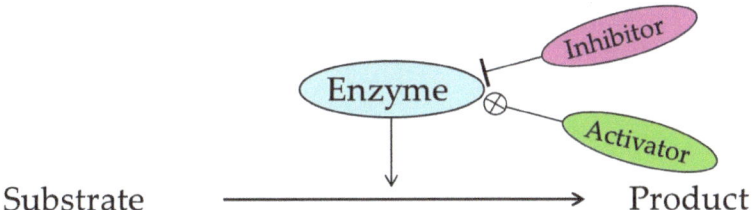

Figure 2: Generalised reaction scheme of an enzyme-catalysed chemical reaction leading to the conversion of a substrate to a product.

1.2 Application of enzyme assays

Enzyme assays are commonly used for two purposes, namely 1) to understand the properties of an enzyme or 2) for bioanalytical applications to identify/determine the concentration of other molecules or the amount of active enzyme present in a preparation.

Among the *properties of an enzyme*, we want to know how fast it is, how strong the enzyme binds the substrate, how specific it is for the substrate, or if any inhibitors exist and how inhibitors modify the enzyme properties. These questions are important in the area of basic science in order to understand how particular enzymes work, how they interact with substrates/inhibitors and the role of enzymes in pathways of metabolism or signal transduction. In biotechnology enzymes are used to make products or to degrade waste, so it is important to know enzyme properties in order to choose the best enzyme for the required application. In biomedical sciences enzyme properties must be determined to identify the causes of illness, which could be the reduced or increased activity of an enzyme or a complete lack of it. In the drug discovery process enzyme properties are determined to identify new inhibitors or activators that may find their way into novel medicines. In agriculture enzyme inhbitors find application as pesticides or herbicides. For example, the herbicide glyphosate is a competitive inhibitor of the enzyme 5-enolpyruvylshikimate-3-phosphate synthase.

An example of a biomedical application is the test for Gaucher's disease. This lipid storage disease is caused by a reduced activity of the enzyme β-glucosylceramidase, and this is tested in a leukocyte homgenisate obtained from patients as illustrated in Figure 3.

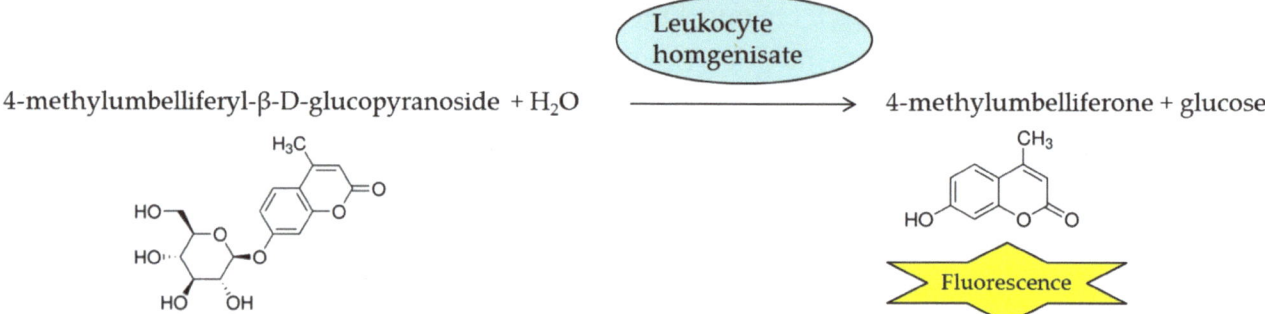

Figure 3: The enzyme assay testing for Gaucher's disease. The leukocyte homogenisate contains among other enzymes β-glucosylceramidase. Gaucher's disease is present, if less than 8.7 nmol product per hour per mg protein is produced.

This example in Figure 3 illustrates two further points of enzyme assays: 1) some enzymes are very specific for their substrate, so a body fluid can be used without further purification, and 2) often an artificial substrate is used that generates a measurable signal.

Bioanalytical applications of enzyme assays are important in answering basic science questions, for example to establish the concentration of other molecules in metabolic pathways or the concentration of proteins in particular cellular states using enzymes linked to antibodies. In biomedical science enzymes may be applied for analytical purposes to test glucose levels, the presence of antibodies or pregnancy. A typical bioanalytical application in biotechnology is to test of how much active enzyme is present in a preparation made by an enzyme producing company. More active preparations can be sold at a higher price. An example for a biomedical application is a test for blood glucose that is carried out with a combination of two enzymes:

1 Introduction

$$\text{glucose} + \text{ATP} \xrightarrow{\text{hexokinase}} \text{glucose-6-phosphate} + \text{ADP}$$

$$\text{glucose-6-phosphate} + \text{NAD}^+ \xrightarrow{\text{glucose-6-phosphate dehydrogenase}} \text{gluconate-6-phosphate} + \text{NADH} + \text{H}^+$$

The reason for using two enzymes is to obtain a product that yields a measurable signal, NADH in this case that can be measure via the absorption of ultraviolet light at a wavelength of 340 nm.

Enzyme linked immunosorbent assays (ELISA) involve enzymes covalently attached to antibodies. Antibodies are protein molecules that can be produced with a high binding specificity towards other molecules called antigens. In Figure 4 the antigen is detected with a specific antibody, while the antibody itself is recognised with another enzyme-linked antibody (part A) or the enzyme is linked to another antigen-specific antibody (part B). The enzyme often produces a colour reaction, for example horseradish peroxidase generates a blue precipitate from the substrate tetramethylbenzidine:

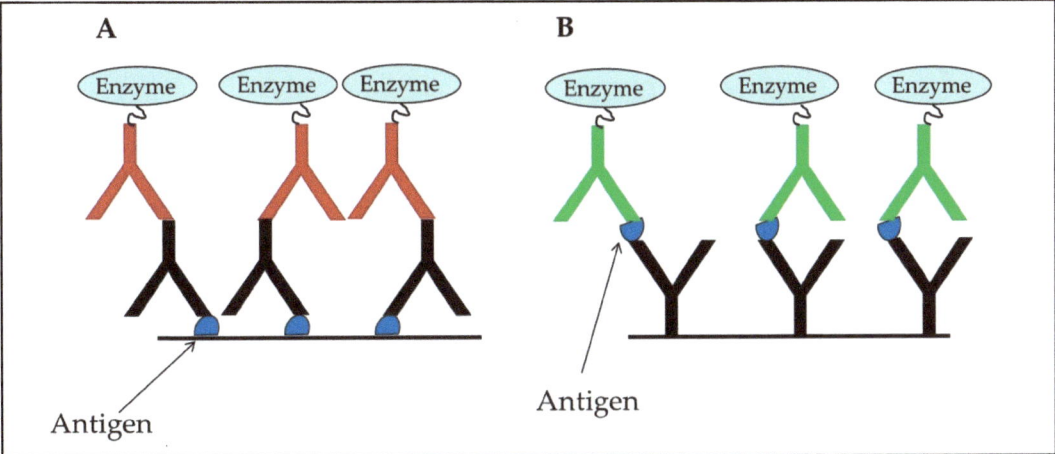

Figure 4: Principle of an ELISA. The antigen molecule is detected with the specific antibody shown in black. Antibodies are large Y-shaped protein molecules; the two arms of the Y bind with high specificity to the antigen. The stem of Y is invariable and can be detected by other enzyme-linked antibodies as in part A, or a sandwich-type ELISA can be performed as shown in part B. The enzyme itself is attached to an antibody and facilitates a reaction that can be detected/ measured.

Enzyme assays may be carried out continuously or discontinuously (Wilson & Walker, 2010). In a *continuous* assay the progress of the reaction is monitored as the reaction proceeds, while in a *discontinuous* assay the reaction is stopped at a certain time and then the concentration of products or substrates is measured. If a complete analysis is required from a discontinuous assay, several reactions must be set up and stopped at various time points. If a detailed kinetic analysis of the enzyme is not required, for example in bioanalytical applications, it is often sufficient to take two measurements at fixed time points. Many types of analytical chemistry techniques can be used to monitor enzyme assays, the preferred ones are UV-VIS spectrophotometry (possible if substances are involved that absorb UV- or visible light), fluorescence (if any of the substances involved shows fluorescence) or chemiluminescence, that is the spontaneous emission of light.

The different time scales for measuring enzyme kinetics are summarised in Figure 5.

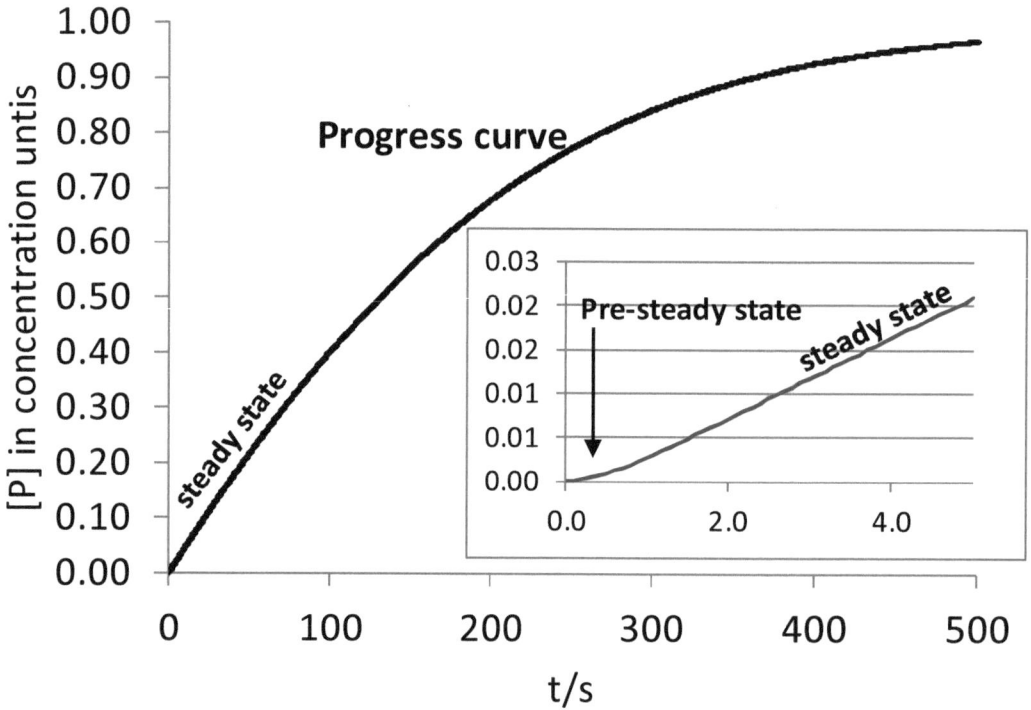

Figure 5: A progress curve showing the concentration of product in dependence of time for a typical enzyme-catalysed reaction converting substrate into product. At the beginning the substrate concentration is 1.0 concentration units and the product concentration increases linearly with time. As the substrate is depleted, a curved relationship becomes apparent. The inset shows the beginning of the reaction at a much higher time resolution. In the first milliseconds there is some delay in the formation of product before the steady-state is reached. This time range is called the pre-steady state. The cuve was calculated with COPASI (Hoops et al, 2006)

1 Introduction

For an enzyme catalysed reaction converting a substrate to product at high substrate concentration there is a linear increase in product concentration with time. This is referred to as the *steady-state*, which may confuse some students as the product concentration keeps increasing. Steady-state refers to the concentration of an intermediate that is almost constant. You may remember at this point that in the steady-state the product concentration increases steadily (= linearly) with time. Most measurements in enzyme kinetics are performed during this steady-state time frame. A requirement for steady-state measurements is that the substrate concentration is much higher than the enzyme concentration. These measurements are also referred to as initial rate measurements or kinetics, which may add to another confusion. The inset of Figure 5 shows that the steady-state is not absolutely initial at the start of the reaction. There is a short delay in the time range below one second before the steady-state is reached. This time range is called the pre-steady-state.

Initial rate or steady-state enzyme kinetics can be performed often with simple equipment, such as pipetting substrate and enzyme together and observing the reaction progress either continuously or discontinuously. If pre-steady state kinetics is to be investigated, the mixing of reagents and placement into the measurement device become limiting factors. Realistically, pipetting an enzyme into a cuvette, mixing and placing it into a spectrophotometer takes at least five seconds, during which no measurement can be taken. This is called the *dead-time*. Pre-steady state kinetics requires the dead-time to be reduced to 0.01 to 0.001 seconds, dependent on how fast the enzyme is. The special techniques for pre-steady state kinetics are briefly described in the following paragraph.

Box 1: Absorbance and calibration graphs

UV-VIS spectrophotometry measures the absorption A of ultraviolett or visible light by a clear solution. If the solution is cloudy due to small particles, such as a precipitate or cells, light passing through the sample is further reduced by turbidity.

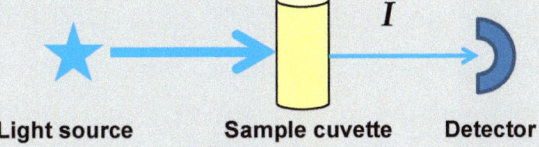

| Light source | Sample cuvette | Detector |

The absorbance by the sample reduces the light intensity I measured at the detector compared to light intensity I_0 for a 'blank'. The absorbance is then defined as

$$A = -\lg(I/I_0) \quad (\lg: \log_{10})$$

Note that this is a practical definition of absorbance with the light intensity I_0 determined via a 'blank' sample. In physics I_0 is considered as the incident light hitting the sample before any absorbance occurs, but due to unpredictable reflections of light on the cuvette containing the sample it is difficult to determine. In practice a blank, that is a buffer solution containing all components of the sample, except the absorbing substance, is used. The absorbance A is related to concentration c by the Beer-Lambert law:

$$A = \varepsilon \, c \, d$$

With ε: molar absorption coeffcient (in dm^3 mol^{-1} cm^{-1}), d: the pathlength or inner diameter of the cuvette and c: concentration (in mol dm^{-3}). In theory, if the absorption coeffcient is known for the substance, the concentration could be calculated from the measured absorbance, but in practice a calibration graph is used as shown in the example below:

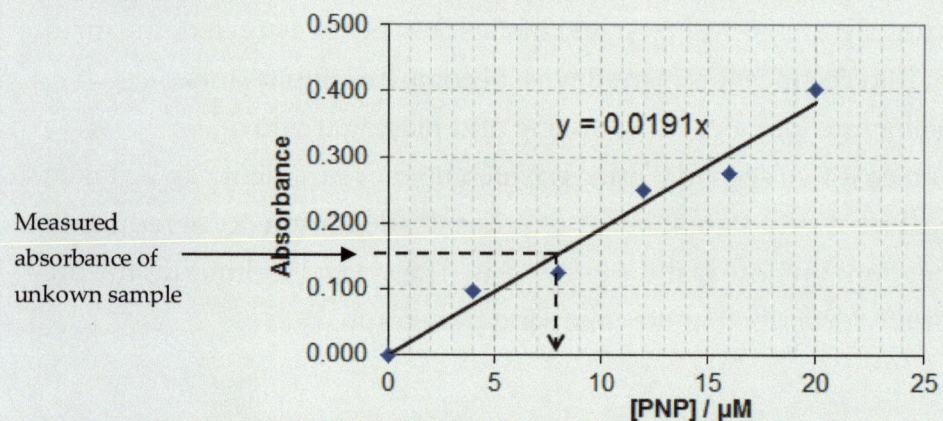

A series of samples of *known* concentration are measured and from this a calibration graph is drawn. If the blank was prepared correctly, the graph starts from the origin. The unknown concentration of samples can then be determined from the absorbance as indicated on the graph, as long as the measured absorbance is within the calibration range. Calibration graphs can be used even if the Beer-Lambert law does not apply; in that case a curved relationship between absorbance and concentration may occur.

1.3.1 Pre-steady state kinetic measurements

The key to pre-steady state enzyme kinetics is to reduce the dead time of the system. This can be achieved by flow systems illustrated in Figure 6. Two or more reagents are injected via syringes into a mixing chamber and then the progress of the reaction is monitored along a reaction tube in the continuous flow system. The distance from the mixing chamber is proportional to the reaction time. A dye solution can be used to calibrate the system. Since biological reagents are expensive, stopped-flow systems have been developed that minimise the wastage of reagents. A stopped-flow system involves a stopping syringe that creates an opposing action and stops the flow after a short time. These systems can reduce the dead-time to 0.001 seconds (1 ms) or even lower with advanced mixing devices.

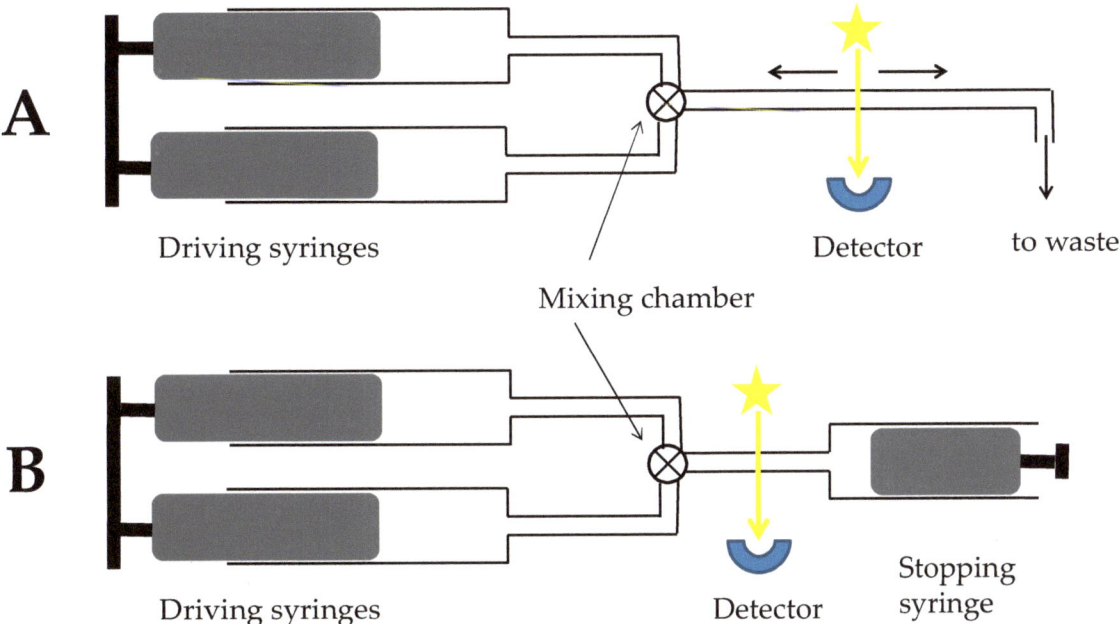

Figure 6: (A) A continuous flow system for mixing two reagents and measuring the reaction progress with spectrophotometry. The spectrophotometer (or the reaction tube) is mobile and the reaction time is proportional to the distance from the mixing chamber. **(B)** A stopped-flow system with a stationary spectrophotometer. The stopped-flow system consumes less reagents.

Relaxation kinetics, devloped by Manfred Eigen, Ronald Norrish and George Porter (Nobel Prize in chemistry, 1967) avoid the problem of mixing but start with a well-mixed reaction system in equilibrium. Then an external perturbation is applied, for example a sudden temperature change. Since the equilibrium of chemical reactions depends on temperature, the equilibrium concentrations of reactants and products will be different at a different temperature. The time it takes for the system to relax to the new equilibrium concentrations is measured. This principle is illustrated in Figure 7. For the simple reaction example in Figure 7, the relaxation x follows the time course $x = x_0 \exp(-t/(k_1+k_{-1}))$, thus the value of rate constants can be obtained in this way. For the details of rate constants see section 2.2.

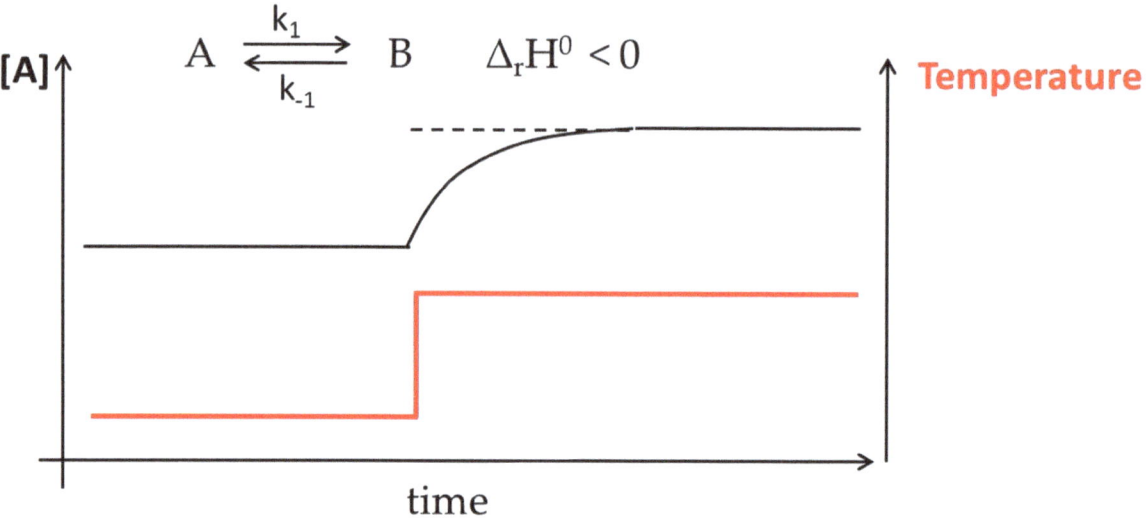

Figure 7: A temperature jump relaxation kinetics experiment for an exothermic reaction. An increase in temperature leads to an increase of the equilibrium concentration of A (dashed line). The relaxation time it takes for the system to reach the new equilibrium is measured.

The dead-time of the system depends on the time it takes to apply the perturbation. For a temperature jump with a pulsed infrared laser dead-times in the range of microseconds ($1 \cdot 10^{-6}$ s) can be achieved. In addition to a temperature jump, pressure jump or an electric field jump is used.

Key points 1 Introduction

- Enzymes are biological catalysts that speed up chemical reactions.
- The majority of enzymes are protein molecules.
- Enzyme assays are laboratory methods that measure enzymatic activity (= a chemical reaction catalysed by an enzyme).
- Enzyme assays may be used to determine the properties of an enzyme.
- Enzyme assays may be used for a bioanalytical purpose: to determine the concentration or presence of a molecule.
- For enzyme assays (under steady-state conditions) it is important that the substrate concentration is much higher than the enzyme concentration.
- Fast kinetic methods such as stopped-flow or relaxation kinetics are required for measuring reaction times in the millisecond range.

Reading for chapter 1

Hoops S, Sahle S, Gauges R et al (2006) COPASI: A COmplex PAthway SImulator, *Bioinformatics* **22:** 3067-3074

Wilson K, Walker J (2010) Principles and Techniques of Biochemistry and Molecular Biology. 7th edition, chapter 15.3, pp 602-611. Cambridge: Cambridge University Press

2 CHEMICAL KINETICS

Kinetics is an area of physical chemistry that is concerned with the rates of chemical reactions (Atkins & de Paula, 2006; Voet & Voet, 2011). From everyday life experiences we might be more familiar with the kinetics of spatial motion, such as running along a path or driving a car. It is instructive to consider an example from the everyday experience of spatial motion kinetics first, before moving on to chemical kinetics. If you are driving a car from X to Z you are covering a certain distance starting at the time t = 0 seconds to t = 360 seconds (for a 10 minutes journey). A plot of the distance travelled against time may look like the curve shown in Figure 8A.

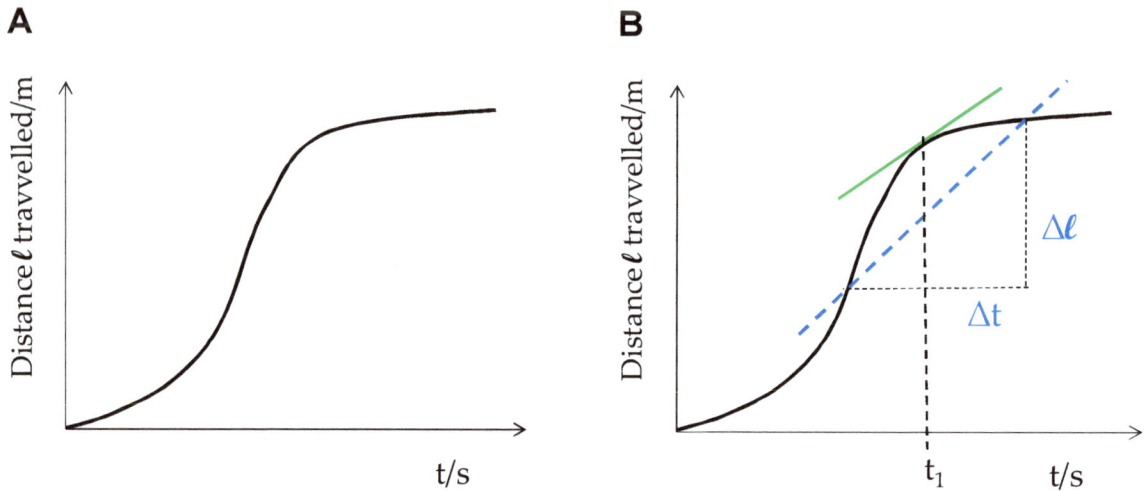

Figure 8: A) The distance travelled during a car journey against time t. B) The average velocity between two time points is the gradient of the dashed blue line $\Delta\ell/\Delta t$, and the instantaneous velocity is the gradient of the green line $d\ell/dt$, a tangent to the curve.

At the beginning the distance travelled increases slowly, then we get onto a fast road and finally we slow down, when we approach our destination. Kinetics is concerned with the velocity and from the distance-time curve we may obtain an average velocity between two points as $\Delta\ell/\Delta t$ shown in Figure 8B by the dashed line. The instantaneous velocity at a time point t_1, that is displayed on the speedometer of the car, is obtained by decreasing the time interval Δt to an infinitely small time interval written as dt and the distance travelled during dt becomes very small as well indicated by $d\ell$, thus the instantaneous velocity is given by $d\ell/dt$. This is known in mathematics as the first derivative of distance with respect to time and part of the branch of calculus in mathematics. The units of this spatial velocity are distance/time, for example m/s or km/h.

In chemistry we are not concerned with distances but with *concentrations* of chemical species. Concentration specifies the amount of a molecule per volume and normally refers to molar concentrations units, such as mol/L (abbreviated as M) or µmol/L (abbreviated as µM). In chemical kinetics we are concerned with how concentrations change over time and everything mentioned above with regards to distance may be replaced by concentration.

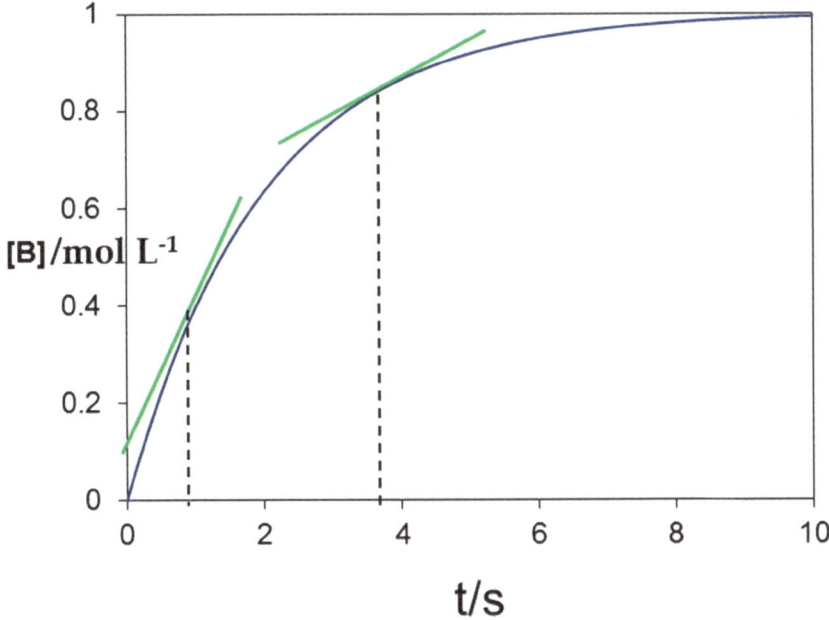

Figure 9: The change of concentration of species B with time for a chemical reaction A ⟶ B. The reaction rate at two different time points (dashed lines) is the gradient of the tangents at the curve.

For the example shown in Figure 9 the velocity v of the chemical reaction is given as the change in concentration of B over time as $v = d[B]/dt$. Note that concentrations are always written in square brackets [], also instead of velocity of a chemical reaction we may use the term *reaction rate*. The units of the reaction rate are concentration/time, for example M/s (speak: "molar per second") (also written as: $M\ s^{-1}$, or $mol\ L^{-1}\ s^{-1}$).

2.1 Definition of the reaction rate

The reaction rate v is defined as the change of concentration per time. It is positive for products and negative for reactants:

For A ⟶ B

$v = d[B]/dt = -d[A]/dt$

Chemical reactions may have different stoichiometries, so we need to take the stoichiometric coefficient into account, for example consider the reaction

2 A ⟶ 3 B, which may be written as 0 = 3 B – 2 A, the reaction rate is

$v = 1/3 \, d[B]/dt = -1/2 \, d[A]/dt$

In general chemical reactions involving various species J with the stoichiometric coefficients v_J (greek letter v) can be written as:

$0 = \sum v_J J$ (0 = 3 B – 2 A in the example above)

The reaction rate is then given as $1/v_J \, d[J]/dt$.

2.2 Rate law and reaction mechanism

2.2.1 Order of rate laws

The rate law specifies how the reaction rate depends on the concentration of a chemical species. Examples of rate laws are:

$-d[A]/dt = k\,[A]$ first order rate law (note that $[A] = [A]^1$)

$-d[A]/dt = k\,[A]^2$ second order rate law

$-d[A]/dt = k$ zero order rate law

$-d[A]/dt = k\,[A]\,[B]$ second order rate law (first order in [A] and first order in [B])

In mathematics a rate law as shown above is called an Ordinary Differential Equation (ODE).

The reaction rate may be proportional to the concentration of a species that means the reaction rate depends on the concentration of A in a linear fashion as in the first example above. This is called a first order rate law. The proportionality constant k is called the rate constant.

Sometimes the reaction rate depends on the concentration of A squared, then we have a second order rate law. The units of the rate constant depend on the order of the rate laws; it can be worked out by solving the equation of the rate law for k, demonstrated here for the last example:

$-d[A]/dt = k\,[A]\,[B]$

$$M/s = \{k\} \, M \, M$$

$$\Leftrightarrow \quad \{k\} = 1/(M\,s) \text{ or } M^{-1} s^{-1}$$

The rate law must be established experimentally and its order is related to the *reaction mechanism*. This is one of the reasons why chemical kinetic studies are performed; they are able to confirm or reject a postulated reaction mechanism. A reaction mechanism specifies the steps of elementary reaction; it reveals how a chemical reaction proceeds in detail step by step. Even an apparently simple reaction may proceed in several elementary reaction steps, e.g. the decomposition of O_3 (ozone): $2\,O_3 \longrightarrow 3\,O_2$ (stoichiometric reaction)

We could imagine that two ozone molecules collide and form three oxygen molecules in one step. However, the reaction mechanism is more complex:

$$O_3 \longrightarrow O_2 + O$$

$$O_3 + O \longrightarrow 2\,O_2$$

If we know the reaction mechanism, we can write down the rate law for each elementary reaction as explained in the following paragraph.

2.2.2 Relationship between rate laws and reaction mechanisms

Consider the following elementary reaction:

$$A \xrightarrow{k} B$$

This is a *uni*molecular reaction as the elementary step involves *one* molecule A that reacts to B. The rate law for this unimolecular reaction is:

$$-d[A]/dt = k\,[A] \quad \text{(a first order rate law)}$$

The rate law seems plausible, since the higher the concentration of A the higher the expected reaction rate. Consider the following elementary reaction:

$$A + B \xrightarrow{k} P$$

This is a *bi*molecular reaction as the elementary step involves two molecules A and B that must collide to form the product P. In that case it is plausible that the reaction rate should depend on both the concentration of A and the concentration of B. Therefore, the rate law for this bimolecular reaction is:

$$-d[A]/dt = k\,[A]\,[B] \quad \text{(a second order rate law)}$$

Consider the following elementary reaction:

$$3\,A \xrightarrow{k} P$$

This is a *tri*molecular reaction as the elementary step involves three molecules of A that must collide to form the product P. Trimolecular reactions occur very seldom in practice, as it is highly improbably that three molecules collide at the same time. The rate law for this reaction is:

$$-d[A]/dt = k\,[A]^3 \quad \text{(a third order rate law)}$$

As we can see from these examples, the molecularity of the elementary reaction determines the form of the rate law. Note the difference between *order* and *molecularity*; the order is obtained experimentally, while the molecularity refers to an elementary reaction of a reaction mechanism.

Pseudo-first order reactions:

Bimolecular reactions may follow a first order rate law, if the concentration of one of the reactants is very high so that it does not change significantly during the course of the reaction. This is typically the case for reactions with water:

$$A + H_2O \xrightarrow{k} B$$

The rate law for this (elementary) reaction is:

$$-d[A]/dt = k\,[A]\,[H_2O] = k'\,[A]$$

The concentration of water (at T = 298 K) is 55.3 mol/L which means that it practically does not change during most reactions. Thus the concentration of water becomes part of a new 'pseudo'-first order rate constant $k' = k\,[H_2O]$. Reactions of this type are known as *pseudo-first order* reactions. When bimolecular reactions are investigated experimentally the concentration of one of the reactants is often chosen so high that it practically does not change. This simplifies the kinetic analysis of bimolecular reactions with pseudo-first order rate laws.

2.3 Integrating the rate law

The rate law provides a mathematical expression for the reaction rate or reaction velocity; this mathematical expression is called an Ordinary Differential Equation (ODE). Often it is useful to know how the concentration changes with time, i.e. what is the concentration at a certain point in time. This involves the mathematical procedure of integrating the ODE, also referred to as *solving* the ODE. The integration can be done explicit, leading to a formula that specifies the concentration at each point in time. This is the preferred and most general form of integrating the rate law; unfortunately it is only possible for simple rate laws. Another method of integration is *numerical integration*, which leads to a series of numbers specifying the concentrations at various time points. Another method of obtaining the time dependent concentrations is molecular simulation. A molecular simulation takes into account individual molecules and the collisions between them. Sometimes numerical integration of rate laws is incorrectly referred to as simulation. As an example to demonstrate the results of explicit and numerical integration we consider the elementary reaction,

$$A \xrightarrow{k} B$$

which was described in section 2.2.2. The results of explicit integration (see Box 2) and numerical integration (see Box 3) are shown in Table 1.

Table 1: Comparison between explicit and numerical integration for the reaction A → B. The concentration of A at the start was set as 1 mol/L and the rate constant as k = 0.5 s^{-1}. For the numerical integration a time step of Δt = 0.5 s was chosen.

Time/s	Explicit: $[B]_t = [A]_0(1-e^{-kt})/(\text{mol L}^{-1})$	Numerical integration: $[B]/(\text{mol L}^{-1})$
0	0	0
1	0.393	0.438
2	0.632	0.684
3	0.777	0.822
4	0.865	0.900
5	0.918	0.944
6	0.950	0.968

The explicit integration is accurate, while the accuracy of numerical integration depends on the time step and the algorithm (method) chosen for numerical integration. For the example shown in Table 1 the simple Euler method was chosen with a time step of 0.5 s. A more advanced algorithm for solving ODEs is the Runge-Kutta method. In Figure 10 the numerical and explicit solutions are compared.

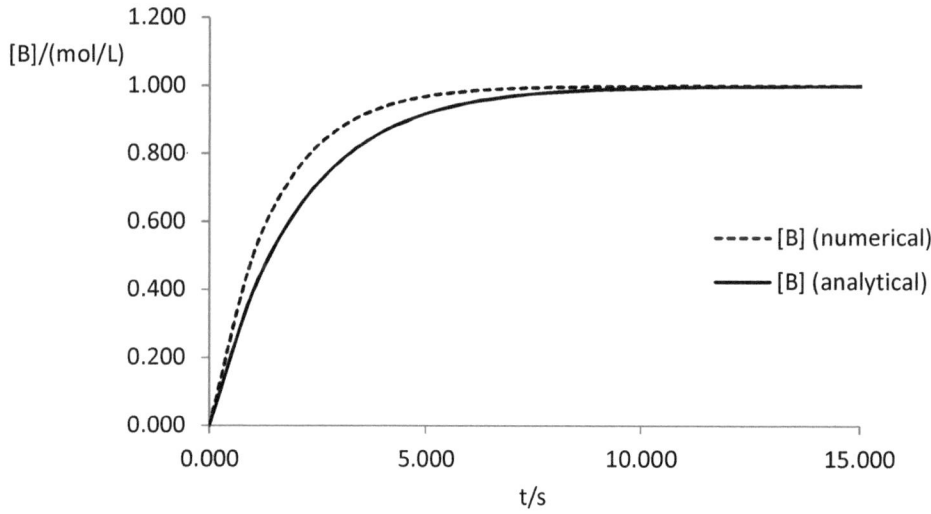

Figure 10: Comparison of explicit and numerical solution of the ODE for a first order reaction.

Box 2: Step by step example of the explicit solution (integration) of an ordinary differential equation (ODE).

$$\frac{d[B]}{dt} = -\frac{d[A]}{dt} = k[A] \qquad \text{this is the rate law}$$

$$d[A]/[A] = -k\,dt \qquad \text{rearrange, 'separation of variables'}$$

$$\int_{A_0}^{A_t} \frac{1}{[A]} d[A] = -k \int_{t=0}^{t} dt \qquad \text{integrate, and set boundary conditions, at t=0 we have}$$

the concentration A_0, and at time t we have the concentration A_t

$$\ln[A]\,\Big|_{A_0}^{A_t} = -kt\,\Big|_{0}^{t} \qquad \text{find the antiderivatives of '1/[A]': ln [A] and on the}$$

right hand side the antiderivative of '1 dt': t. At the same time we keep hold of the boundary conditions by noting them along the vertical line. Web-sites can help with integration, such as the Wolfram integral calculator: http://www.wolframalpha.com/calculators/integral-calculator.

$$\ln A_t - \ln A_0 = -k(t-0) \qquad \text{insert boundary conditions}$$

$$\ln \frac{A_t}{A_0} = -kt \qquad \text{rearrange (use: ln a – ln b = ln (a/b))}$$

$$\frac{A_t}{A_0} = \exp(-kt) \qquad \text{apply exp-function to both sides (use: exp(ln x) = x)}$$

$$A_t = A_0 \exp(-kt) \quad \text{(a)} \qquad \text{rearrange (this is the final equation for } A_t\text{)}$$

In order to find the equation for the concentration of B, we need to apply the principle of the conservation of mass, namely at any point in time the concentrations of A and B together are equal to the starting concentration A_0:

$$A_t + B_t = A_0$$

$$B_t = A_0 - A_t \qquad \text{rearrange and substitute equation (a) for } A_t:$$

$$B_t = A_0 - A_0 \exp(-kt) = A_0(1 - \exp(-kt)) \quad \text{(this is the final equation for } B_t\text{)}$$

Box 3: Example of a numerical solution of an ODE with the Euler method

The numerical solution of the ODE with the Euler method is based directly on the rate law, for the example A → B:

d[B]/dt = k [A]

d[B] = k [A] dt

The infinitely small intervals d[B] and dt in the differential equation may be approximated by:

Δ[B] = k [A] Δt, which leads directly to the numerical integration (with starting conditions A_0 = 1, B_0 = 0 and k = 0.5 s^{-1} as in Table 1). And we need to choose a time interval Δt = 0.5 s (the smaller the time interval, the more accurate the numerical integration will be).

Step 1: ΔB = 0.5 s^{-1} · 1 M · 0.5 s = 0.25 M   =>   [B] (at t=0.5 s) = 0.25 M
           k   [A]   Δt   [A] (at t=0.5 s) = 1 - 0.25 = 0.75 M

Step 2: ΔB = 0.5 s^{-1} · 0.75 M · 0.5 s = 0.1875 M => [B] (at t=1.0 s) = 0.25 M + 0.1875 M
                                                                     = 0.4375 M
                                                        [A] (at t=1.0 s) = 0.5625 M

Step 3: ΔB = 0.5 s^{-1} · 0.5625 M · 0.5 s = 0.141 M =>[B] (at t=1.5 s) = 0.579 M
                                                        [A] (at t=1.5 s) = 0.421 M

And so on ...

Note that the time step Δt determines the accuracy of the numerical solution, in particular Δt should be smaller than the inverse value of the rate constant, hence the limited accuracy in our example.

2.4 Rate laws for complex reaction mechanisms

For complex multi-step reaction mechanisms the rate laws are a combination of the individual steps. Consider the example

$$A \underset{k_{-1}}{\overset{k_1}{\rightleftarrows}} B$$

This may be written as individual reactions:

$$A \xrightarrow{k_1} B$$
$$B \xrightarrow{k_{-1}} A$$

The rate law is the combination (sum) of the forward reaction (described by k_1) and the backward reaction (described by k_{-1}):

$$\frac{d[A]}{dt} = -k_1[A] + k_{-1}[B]$$

     forward + backward

The sign and multiplier in front of the rate constant follows the rules about the definition of the reaction rate in section 2.1 and is given by $1/\nu_J$ (with ν_J as the stoichiometric coefficient). For the example above, the stoichiometric coefficient of A is -1 (the reaction can be written as $0 = B - A$) and the stoichiometric coefficient of B is +1. Alternatively, we may say that A *disappears* to B with k_1, hence -1, and B is *formed from* A with k_{-1}, hence +1 in front of the concentration of B. The reaction rate with regards to B is:

$$\frac{d[B]}{dt} = k_1[A] - k_{-1}[B]$$

The previous two differential equations describe the kinetics of the reaction completely. If more than one ODEs is necessary to describe the kinetics we speak of a system of ODEs. Consider another set of elementary reactions, which is a typical mechanism for enzyme catalysis:

$$E + S \underset{k_{-1}}{\overset{k_1}{\rightleftarrows}} ES \xrightarrow{k_2} P + E$$

The kinetics of this reaction is described by a system of ODEs:

$$\frac{d[E]}{dt} = -k_1[E][S] + k_{-1}[ES] + k_2[ES]$$

$$\frac{d[S]}{dt} = -k_1[E][S] + k_{-1}[ES]$$

$$\frac{d[ES]}{dt} = k_1[E][S] - k_{-1}[ES] - k_2[ES]$$

$$\frac{d[P]}{dt} = k_2[ES]$$

Given the principles and examples discussed you should be able to write down the rate laws for any set of elementary reactions (further explanations in Box 4).

Box 4: A practical set of rules to derive rate equations for complex reaction mechanisms.

The following provides a practical set of rules for deriving the rate equation of a molecular species X involved in a complex reaction mechanism. $d[X]/dt =$

(1) the rate constants for all reactions leading towards X contribute with a positive sign.
(2) the rate constants for all reactions leading away from X contribute with a negative sign.
(3) the concentrations of the reactants contribute according to the reaction order.
(4) the factor in front of the concentrations is '1' divided by the stoichiometric number (without sign, as this has been already taken care of by points (1) and (2) above).

$$A \underset{k_2}{\overset{k_1}{\rightleftarrows}} 2X \xrightarrow{k_3} P$$

Applying rule (1) and (2): $d[X]/dt$: $+k_1$, $-k_2$, $-k_3$

Applying rule (3): reaction 1 first order $[A]^1$, reaction 2 second order $[X]^2$, reaction 3 second order $[X]^2$.

Applying rule (4): factors (without signs) are for A: 1, for X: ½ and for P: 1 (but not required).

$$\frac{d[X]}{dt} = k_1[A] - \frac{1}{2}k_2[X]^2 - \frac{1}{2}k_3[X]^2$$

2.5 Compound rate laws

For a stoichiometric reaction rate laws can be determined experimentally by measuring how the reaction rate depends on the concentration of reactants and products. These compound rate laws do not follow the simple principles outlined above for elementary reactions. Compound rate laws may have non-integral orders and depend on concentrations in unexpected ways, such as $[A]^{1/2}$ or $[A]^{-1}$. However, the compound rate law must be compatible with the reaction mechanism and can be derived from the elementary rate laws. For the example of the enzyme catalysis mechanism introduced above, where a substrate S reacts to the product P in presence of an enzyme: S → P, the compound rate law is:

$$v_0 = \frac{a[S]}{b+[S]}$$; with 'a' and 'b' constants that do not depend on the concentration of S or P.

This rate law can be experimentally found and theoretically derived from the reaction mechanism shown above. The derivation of this rate law is shown in section 3.3. Another example is the reaction between two gases:

H₂ + Br₂ ⟶ 2 HBr

The compound rate law is:

$$v = \frac{k_1[H_2][Br_2]^{3/2}}{[Br_2]+k_2[HBr]}$$

Again this rate law was experimentally found and can be derived from a postulated reaction mechanism.

2.6 The rate limiting step

In complex reaction mechanisms as discussed in the previous chapter rate constants may take on any value. If rate constants are very different by factors of ten or more, the slowest reaction determines the overall reaction rate. This reaction step is then called the *rate limiting step*. This is not always, but often the reaction with the lowest rate constant. The situation is similar to a motorway with three lanes, where at one point two lanes are blocked. The overall rate of traffic is only as fast as cars can pass the narrow one-lane part. Consider the following example:

A → B → C with rate constants k_1 and k_2. The concentration of C in dependence of time can be obtained through explicit integration of the system of ODEs:

$$[C]_t = [A]_0 + [A]_0 \left(\frac{k_1 e^{-k_2 t} - k_2 e^{-k_1 t}}{k_2 - k_1} \right)$$

If the first step is rate limiting, that means k_1 is much smaller than k_2 we can simplify the equation above by taking into account that in the denominator $k_2 - k_1 \approx k_2$ and in the numerator $k_1 exp(-k_2\, t)$ is very small compared to $k_2 exp(-k_1\, t)$. Then the equation simplifies to:

$$[C]_t = [A]_0 + [A]_0 \left(\frac{-k_2 e^{-k_1 t}}{k_2} \right) = [A]_0 + [A]_0 \left(-e^{-k_1 t} \right) = [A]_0 \left(1 - e^{-k_1 t} \right)$$

This is the same as the explicit solution to a first order reaction as shown in chapter 2.3, namely the reaction A → C with the rate constant k_1. If the second step is rate limiting we would obtain a similar expression with the rate constant k_2. Complex reaction mechanisms can often be simplified by considering a rate limiting step.

The rate limiting step of complex reaction schemes is *not* always the reaction with the lowest rate constant, in particular if forward and reverse reactions occur. As an example consider the following reaction mechanism:

$$A \underset{100\,s^{-1}}{\overset{1\,s^{-1}}{\rightleftarrows}} I \xrightarrow{10\,s^{-1}} B$$

In this example the second reaction (I → B) is rate limiting. Due to the fast backward reaction (I → A), the concentration of I is always very low, thus the individual reaction rate of the second reaction is smaller than the individual reaction rate of the first forward (A → I) reaction. This is confirmed by the numerical analysis shown in Figure 11.

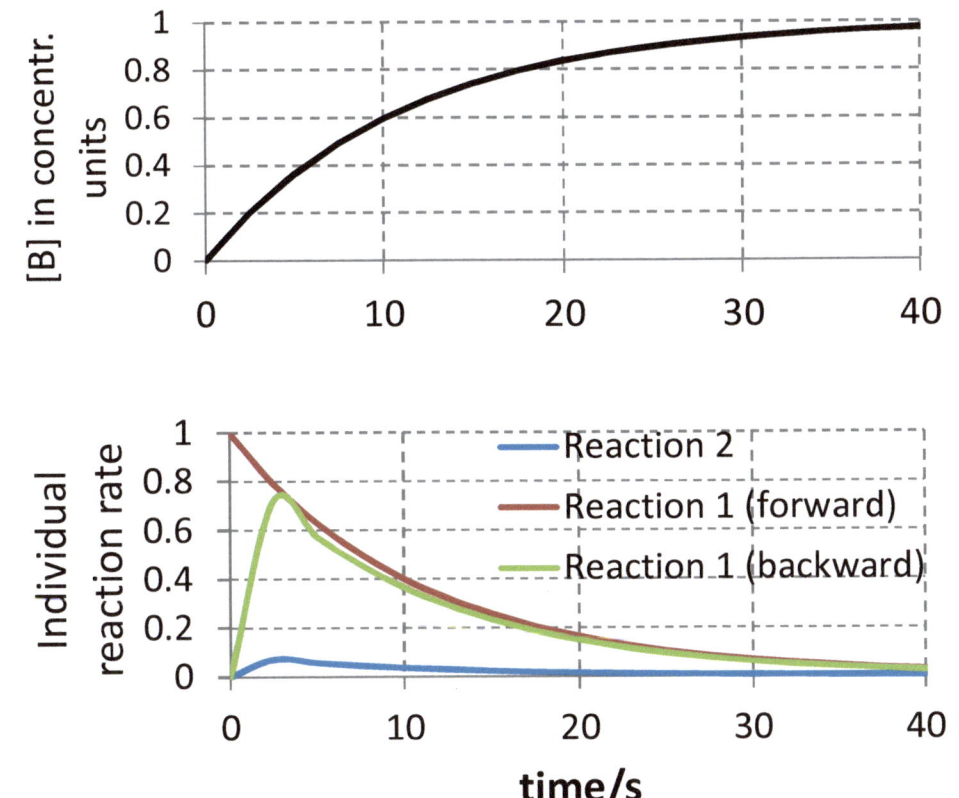

Figure 11: Results of a numerical integration. Top: the concentration of the final product B plotted against time. Bottom: the individual reaction rate for each reaction.

Figure 11 shows that reaction two is always slower than the other reactions. Furthermore, after approximately 2.5 s the forward and backward reaction rates of reaction one are equal. Equal forward and backward reaction rates mean that an equilibrium has established itself. The particular reaction introduced above is an example of a *pre-equilibrium*. The approximation of a fast pre-equilibrium is sometimes used to *simplify* the kinetic analysis of complex reaction schemes. In section 3.3 the steady-state approximation will be introduced as another way of simplifying the kinetic analysis.

2.7 Temperature dependence of reaction rates

The reaction rate increases with increasing temperature; as a rule of thumb raising the temperature by 10 K doubles the reaction rate. The temperature dependence of a rate constant is described by the Arrhenius equation:

$$k = A\exp\left(-\frac{E_A}{RT}\right)$$

with k: rate constant, A: pre-exponential factor (in the units of the rate constant), E_A: activation energy (in J mol^{-1}), R = 8.314 J K^{-1} mol^{-1} and T: temperature in Kelvin.

The pre-exponential factor A is related to the collision between molecules and to internal motions (in a bimolecular reaction) or to internal motions of the molecule only (in a unimolecular reaction). While there are theories that allow calculation of A from molecular properties, it is usually determined through experiments. When the ratio of two rate constants at two different temperatures T_1 and T_2 is considered, A cancels out (see Box 5):

$$\frac{k_2}{k_1} = \exp\left[\frac{E_a}{R}\left(\frac{1}{T_1} - \frac{1}{T_2}\right)\right]$$

The activation energy can be interpreted as an energy barrier that must be overcome before the reaction proceeds to products. The point at the top of this energy barrier is called the *transition state*. The diagram in Figure 12 illustrates the concept of activation energy.

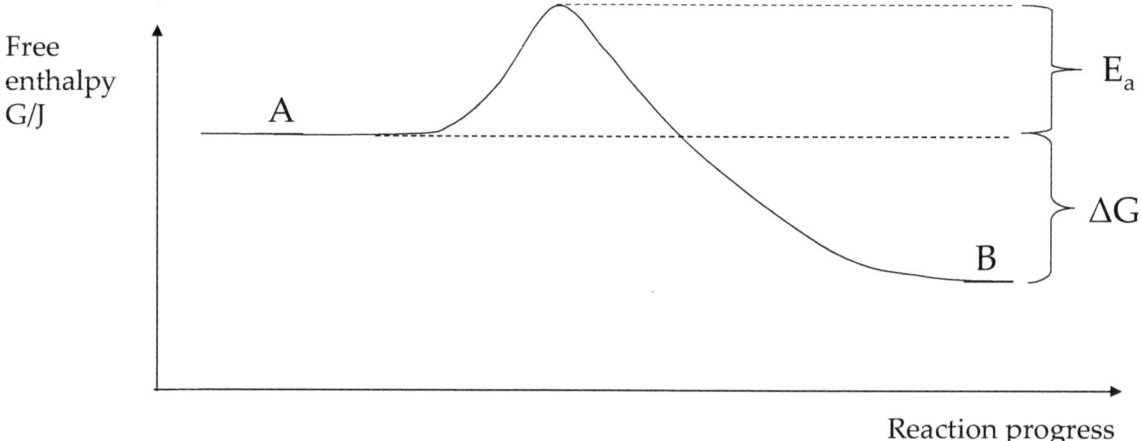

Figure 12: A reaction energy profile for the exergonic reaction A → B. E_a is the activation energy and ΔG is the free enthalpy change of the reaction.

As a reaction proceeds the reactants must overcome this energy barrier. With increasing temperature the thermal energy of the reactants increases, which makes it more likely for them to overcome the energy barrier. This temperature dependence is described by the Arrhenius equation.

The height of the activation barrier is related to the rate constant as
$E_A = (\ln A/k)/RT$, i.e. the lower the rate constant, the higher the activation barrier. Some textbooks state *incorrectly* that for multistep reactions the rate limiting step is always the reaction with the highest activation barrier (see section 2.6).

Box 5: The Arrhenius equation expressed as the ratio of two rate constants.

The ratio of rate constants k_1 at temperature T_1 and k_2 at temperature T_2 can be written as:

$$\frac{k_2}{k_1} = \frac{A\exp\left(-\frac{E_A}{RT_2}\right)}{A\exp\left(-\frac{E_A}{RT_1}\right)} \quad \text{(the pre-exponential factor A cancels out)}$$

$$= \frac{\exp\left(-\frac{E_A}{RT_2}\right)}{\exp\left(-\frac{E_A}{RT_1}\right)} \quad \text{(the exponential is resolved } \exp(A)/\exp(B) = \exp(A-B)\text{)}$$

$$= \exp\left(-\frac{E_A}{RT_2} - \left(-\frac{E_A}{RT_1}\right)\right) \quad \text{(rearrange)}$$

$$= \exp\left(\frac{E_A}{RT_1} - \frac{E_A}{RT_2}\right) \quad \text{(extract common factor } E_A/R\text{)}$$

$$= \exp\left(\frac{E_A}{R}\left(\frac{1}{T_1} - \frac{1}{T_2}\right)\right)$$

2.8 A molecular picture of chemical reactions

Finally we want to ask the question as to the origin of the rate equations. For that we imagine a test tube filled with millions of reactant molecules dissolved in water. The molecules move around in a random fashion (a process that is called diffusion) and occasionally collisions occur. Some of these collisions may not lead to anything, but some collisions lead to a chemical reaction along the pathway of a reaction mechanism. Take as an example the reaction mechanism below:

$$E + S \rightleftharpoons ES \longrightarrow P + E$$

A snapshot of this reaction taken at different times could be like that shown in Figure 13.

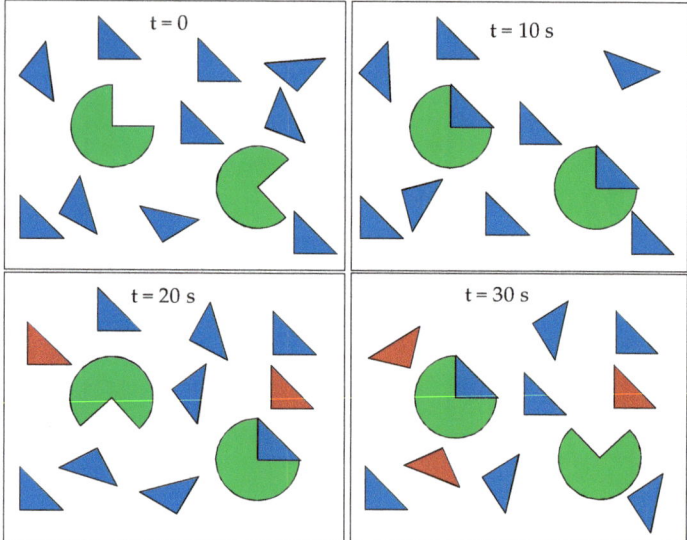

Figure 13: Snapshots of the reaction of molecules in a test tube containing an enzyme (green), substrate (blue) and product (red). The relative sizes of molecules are not accurate, usually enzyme molecules are much larger than substrates, unless the substrate is a protein itself.

The reaction between individual molecules is a random process. There is a certain probability that two molecules E and S form an ES complex and once the ES complex has formed there is a certain probability that it may react to product P or fall apart to E + S again. The rate constants and differential equations that apply to chemical kinetics are averages over millions of these random events. If we were able to look at one molecule at a time, we would observe a stochastic variation of reaction rates but on average the reactions rates would be the same as those described by the differential equations. Nowadays scientists are able observe single enzyme molecules and we will cover this topic briefly the last chapter of the 'Lecture Notes'

Chemical Kinetics

Key points chapter 2 Chemical Kinetics:

- The *reaction rate* is defined as the change of concentration of a chemical substance J per time: $v = 1/v_J \, d[J]/dt$
- The *rate law* specifies the relationship between reaction rate and concentrations of substances. Mathematically the rate law is an ordinary differential equation (ODE)
- The *reaction order* is given by the sum of the exponents of the concentrations, e.g. $v = k[A]^2[B]$ has the order 2+1 = 3.
- A first-order rate law describes a uni-molecular reaction, a second-order rate law describes a bi-molecular reaction (for elementary reactions).
- In order to obtain the concentration of substances over time, the rate law must be *integrated* either numerically or analytically.
- The rate laws for *complex reactions* are obtained by adding the rate laws for elementary reactions.
- Complex reaction mechanisms may be simplified by considering the *rate limiting step*, *pseudo-first order* reactions, *fast pre-equilibria* and the the *steady-state approximation*.
- The Arrhenius equation $k = A \exp(-E_a/(RT))$ describes the *temperature dependence* of rate constants.
- Increasing the temperature by 10 K approximately doubles the reaction rate.
- The temperature dependence is caused by a larger number of molecules that are able to overcome the *activation energy* at higher temperature.
- Chemical kinetics on a molecular level shows a 'stochastic' behaviour. Rate laws describe averages over millions of molecules.

Reading for chapter 2:

Atkins P, de Paula J (2006) Physical Chemistry for the Life Sciences. Section II - The Kinetics of Life, pp 238-339. New York: W.H. Freeman

Voet D, Voet JG (2011) Biochemistry. Chapter 14-1, pp 482-487. New York: Wiley

3 Enzyme Kinetics

3.1 The transition state theory applied to enzymes

As discussed in chapter 2.7, chemical reactions proceed through a *transition state* that is of higher energy than the reactants and the products, thus the reactants must overcome the energy barrier of the transition state called the activation energy. According to the transition state theory the reaction rate depends on the height of the activation energy. Enzymes as biological catalysts speed up the rate of chemical reactions. This can be explained by a reduction of the activation energy by the enzyme. Consider the following example:

A-B $\rightleftharpoons$ A + B

The energy profile of this reaction shown in Figure 14 illustrates the concept of the reduction of activation energy by the enzyme.

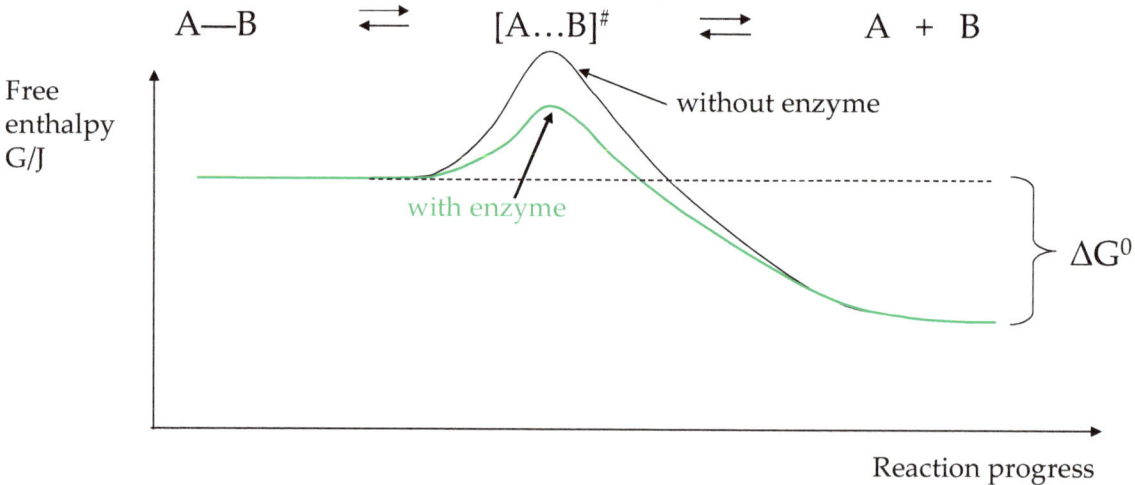

Figure 14: Reaction energy profile of an elementary reaction. The symbol # denotes the transition state.

The consequence of the reduction in activation energy is that more reactant molecules will have the chance to overcome the activation barrier, thus the reaction proceeds faster.

How are enzymes able to reduce the activation energy? This can be explained by the enzyme providing a structural environment that stabilises the transition state or in other words reduces the energy of the transition state. Efficient enzymes should ideally be complementary to the transition state.

3.2 The reaction mechanism of alkaline phosphatase

Alkaline phosphatase is an enzyme that hydrolyses esters of phosphoric acid into the corresponding alcohol and phosphoric acid. The kinetics of enzymes is often studied with model substrates (see chapter 1.2) that yield a colour, so the progress of the reaction can be followed with a spectrophotometer. In case of alkaline phosphatase, the substrate p-nitro-phenylphosphate is often used that is hydrolysed into p-nitro-phenol and hydroxy-phosphate:

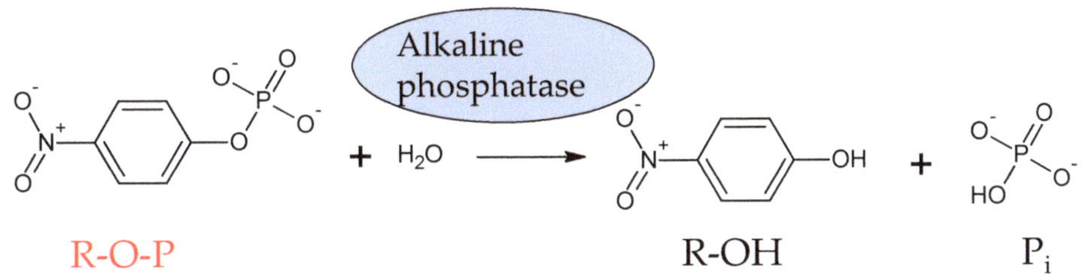

$$R\text{-}O\text{-}P + H_2O \xrightarrow{\text{Alkaline phosphatase}} R\text{-}OH + P_i$$

The mechanism of this reaction is illustrated in the following figures (reviewed in Holtz & Kantrowitz, 1999):

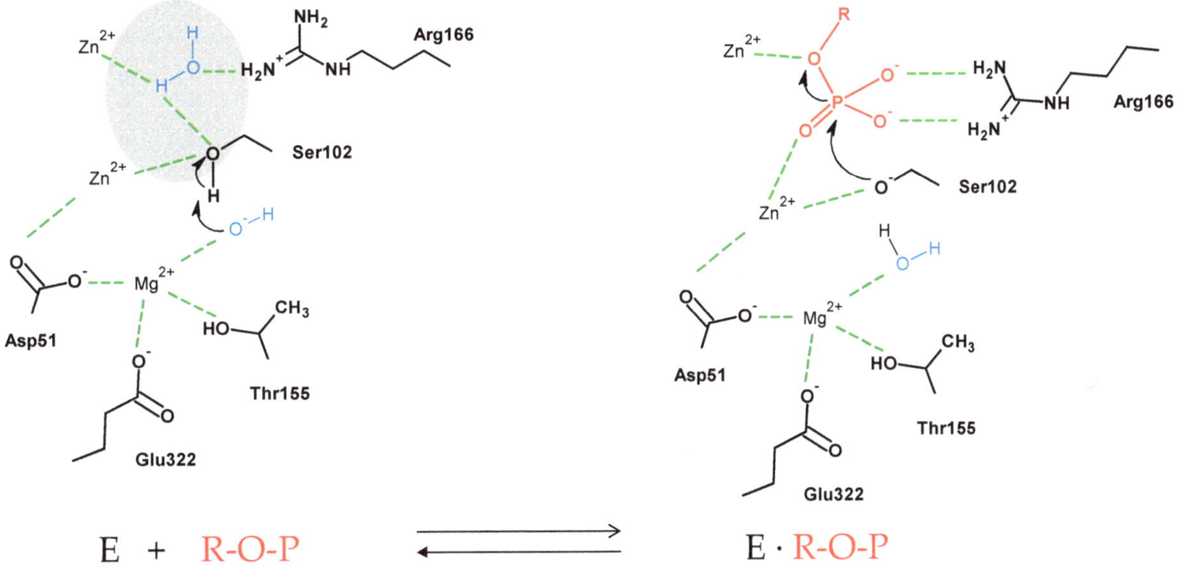

$$E + R\text{-}O\text{-}P \rightleftharpoons E \cdot R\text{-}O\text{-}P$$

Figure 15A: Initially the active site of the enzyme contains one water molecule and hydroxyl-ion (OH-) as well as the two Zn^{2+} and one Mg^{2+} ion as co-factors. Hydroxyl ions are formed under alkaline conditions, which explains why the enzyme prefers an alkaline pH of >10. In the first step the negatively charged substrate replaces the water molecule and associates with the enzyme facilitated by interactions with the positively charged Arg166 and the two Zn^{2+} ions. This illustrates the formation of the enzyme-substrate complex. The hydroxyl-ion extracts a proton from the residue Ser102.

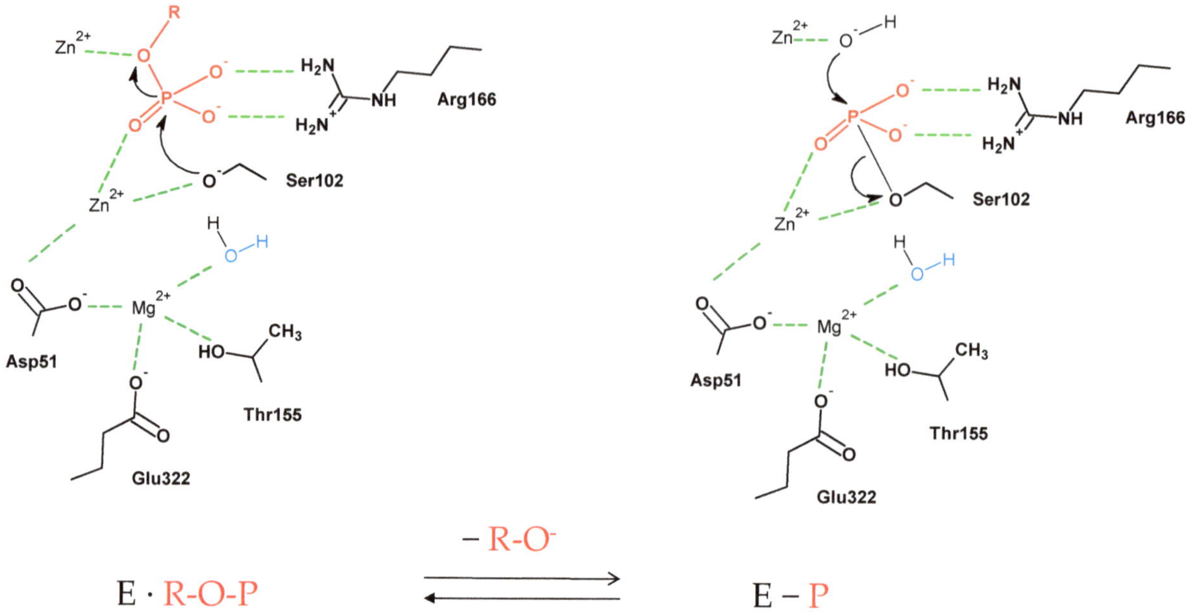

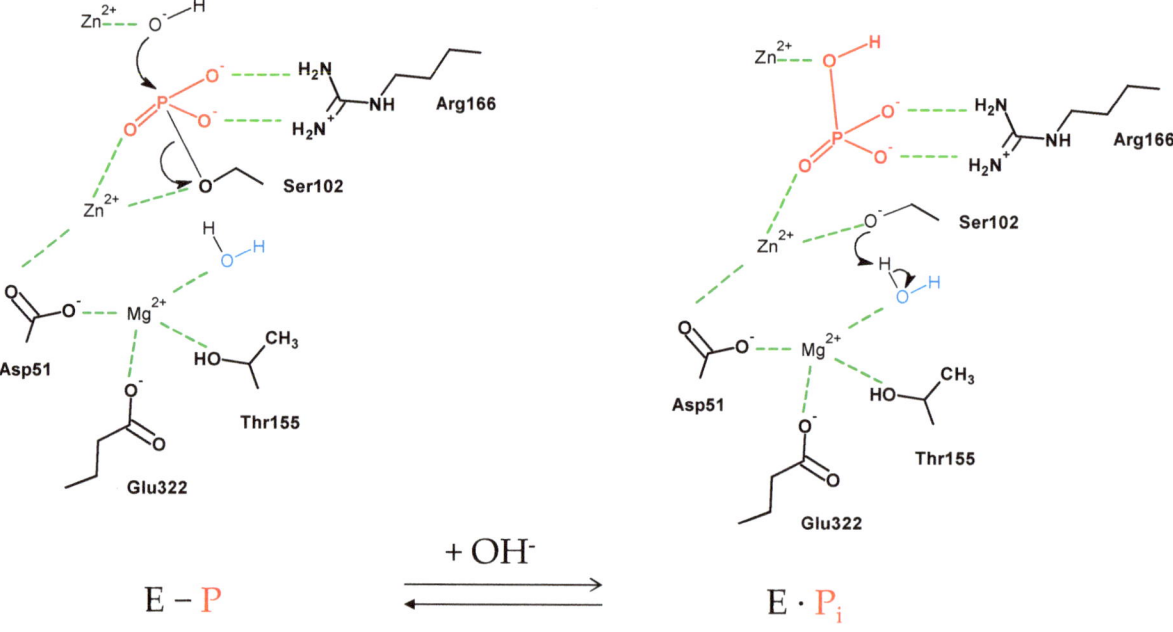

$$E \cdot R\text{-}O\text{-}P \xrightleftharpoons{-R\text{-}O^-} E - P$$

Figure 11B: The deprotonated Ser102 is a strong nucleophile that substitutes the alcohol (p-nitrophenol) of the phosphate in a nucleophilic substition. The alcohol leaves the active site of the enzyme and the phosphate remains covalently bound to Ser102. The top Zn^{2+} ion associates with a hydroxyl-ion in preparation for the next step.

$$E - P \xrightleftharpoons{+OH^-} E \cdot P_i$$

Figure 11C: A hydroxyl ion stabilised by one of the Zn^{2+} cofactors carries out a nucleophilic substitution at the phospho-serine and the bond between the phosphate and Ser102 breaks. The inorganic hydroxy-phosphate remains loosely attached to the enzyme.

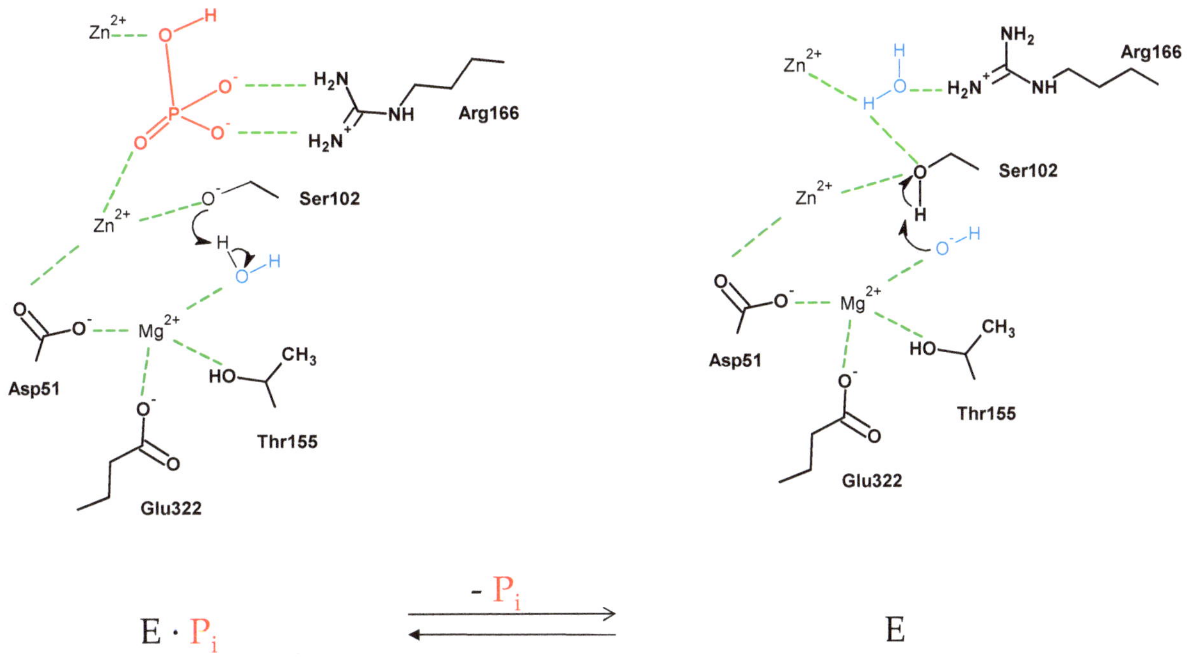

3 Enzyme Kinetics

[Diagram showing two active site structures with Zn²⁺, Mg²⁺, Asp51, Arg166, Ser102, Thr155, Glu322 residues, connected by an arrow labeled "− P_i"]

$$E \cdot P_i \quad \xrightleftharpoons{- P_i} \quad E$$

Figure 11D: In the final step the hydroxy-phosphate becomes replaced by a water molecule and the unchanged active site of the enzyme is available for another round of catalysis. Note that water is available in abundance at a concentration of 55 mol/L.

The reaction mechanism illustrated in Figure 15 shows that simple organic chemistry is going on in the active site of an enzyme. The difference to normal organic chemistry is the precise spatial arrangement of amino acid side chains and cofactors that leads to a substantial increase of the reaction rate as compared to the spontaneous hydrolysis of p-nitro-phenylphosphate in water (without enzyme). The reaction mechanism can be summarised as follows:

$$E + S \; \rightleftharpoons \; E \cdot S \; \rightleftharpoons \; E\text{-}P_2 + P_1 \; \rightleftharpoons \; E \cdot P_2 \; \rightleftharpoons \; E + P_2$$

The formation of the enzyme-substrate complex ES is usually a fast reaction that is in equilibrium. Once products have been formed, the backward reaction, i.e. the formation of ester from the alcohol and phosphate, can be neglected. This simplifies the reaction mechanism to:

$$E + S \; \rightleftharpoons \; E \cdot S \; \longrightarrow \; E\text{-}P_2 + P_1 \; \longrightarrow \; E \cdot P_2 \; \longrightarrow \; E + P_2$$

Finally, if we assume the breaking of the ester bond and formation of P_1 as the *rate limiting* step, the reaction mechanism can be simplified further:

$$E + S \rightleftarrows ES \longrightarrow E + P \quad \text{(with } P = P_1 + P_2\text{)}$$

This is a general mechanism in enzyme kinetics that leads to the widely applied Michaelis-Menten equation. The enzyme reacts with the substrate to form an enzyme-substrate complex ES, which reacts further to to free enzyme and products.

3.3 The Michalis-Menten equation

The simplified mechanism of enzyme kinetics leads to the following system of ODEs (as explained in chapter 2.4):

$$\frac{d[E]}{dt} = -k_1[E][S] + k_{-1}[ES] + k_2[ES]$$

$$\frac{d[S]}{dt} = -k_1[E][S] + k_{-1}[ES]$$

$$\frac{d[ES]}{dt} = k_1[E][S] - k_{-1}[ES] - k_2[ES]$$

$$\frac{d[P]}{dt} = k_2[ES]$$

This system of ODEs cannot be solved explicitly, a numerical solution is shown in Figure 16. It can be seen in Figure 16A that the product concentration [P] increases linearly with time, while the concentration of the enzyme substrate complex [ES] is almost constant.

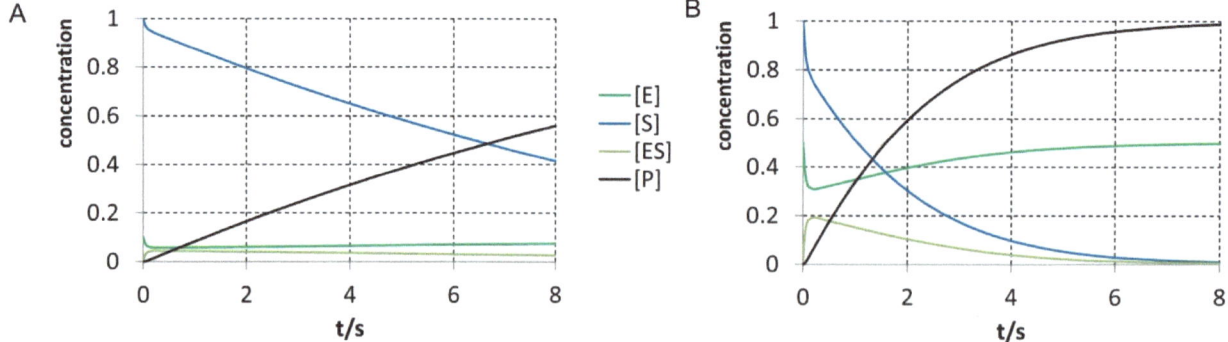

Figure 16: Concentrations of enzyme E, substrate S, enzyme-substrate complex ES and product P obtained from numerical integration of the enzyme kinetics mechanism with $k_1 = 10$ L mol^{-1} s^{-1}, $k_{-1} = 10$ s^{-1} and $k_2 = 2$ s^{-1}. A) With a starting concentration of enzyme of 0.1 (arbitrary units) and substrate of 1. B) With a starting concentration of enzyme of 0.5 (arbitrary units) and substrate of 1.

While in Figure 16A the enzyme concentration was 1/10th of the substrate concentration, in Figure 16B a much higher enzyme concentration of 0.5 (in arbitrary units) was used. In this case the concentrations change in a non-linear fashion with time. In most standard enzyme assays conditions as in Figure 16A are applied. From the graph in Figure 16A we can see that the concentration of the enzyme substrate complex [ES] and the concentration of the enzyme [E] remain at a low constant value. If a property is constant over time, it means that the rate or gradient against time is zero. With this information we can simplify our system of differential equations shown above by setting

$$\left(\frac{d[ES]}{dt}\right)_{t\approx 0} = 0$$

This is called the *steady-state assumption* originally made by Briggs and Haldane in 1925. It is valid for the *initial phase* of the reaction as long as there is much more substrate than enzyme available (see also Figure 5). The duration of this initial phase depends on the enzyme (and substrate concentration); for very fast enzymes such as catalase (hydrogen peroxide decomposition) the initial phase can be very short, while for other enzymes such as alkaline phosphatase the initial phase can last for a couple of minutes. With the steady-state assumption we can derive a simplified expression for the reaction rate v_0 for the initial phase of the enzyme reaction (see Box 6 for details):

$$v_0 = \left(\frac{d[P]}{dt}\right)_{t\approx 0} = \frac{k_2[E]_T[S]}{K_M + [S]} \quad (1a)$$

In this equation $[E]_T$ is the total enzyme concentration and $K_M = (k_{-1} + k_2)/k_1$ is called the Michaelis constant. Equation 1a is known as the Michaelis-Menten equation and it can be simplified further by considering what is the maximum achievable rate of product formation, i.e. the absolute maximum for $d[P]/dt = k_2 [ES]$. This maximum rate would be achieved, when all available enzyme is saturated with substrate, i.e. $[ES] = [E]_T$. With $v_{max} = k_2 [E]_T$ equation 1a becomes:

$$v_0 = \frac{v_{max}[S]}{K_M + [S]} \quad (1b)$$

Equation 1b is the most common form of the Michaelis-Menten equation. Note that in the derivation of equation 1 no integration has been applied, yet it provides the basis for determining the enzyme kinetic parameters v_{max} and K_M by measuring the initial reaction rate v_0 at various substrate concentrations.

The following assumptions are underlying the Michaelis-Menten equation:

- The mechanism shown on page 34,
- and in particular, that the reverse reaction of enzyme with product (E + P → ES) can be neglected.
- The substrate concentration must be much higher than the enzyme concentration.
- The initial phase of the reaction is considered.
- The steady-state must have been reached.

The two last assumptions in this list may seem contradictory. If we were able to measure the first couple of milliseconds of an enzyme reaction, we would notice that the product concentration does increase in a curved fashion as can be seen at times close to zero in Figure 5. In this millisecond time range the Michaelis-Menten equation does not apply. This data can be obtained with advanced experimental techniques for fast kinetics, such as stopped-flow or relaxation methods (temperature-jump, pressure jump, see section 1.3.1).

Box 6: Derivation of the Michaelis-Menten equation

$$\frac{d[P]}{dt} = k_2[ES] \qquad \textbf{(a)}$$

$$\frac{d[ES]}{dt} = k_1[E][S] - k_{-1}[ES] - k_2[ES] \qquad \textbf{(b)}$$

The steady-state assumption means that we can set equation **b** to zero, which allows us to obtain an expression for [ES] that can be substituted into equation **a**.

Equation b: $0 = k_1[E][S] - k_{-1}[ES] - k_2[ES]$ \qquad ($-k_1[E][S]$ and multiply by -1)

$$k_1[E][S] = k_{-1}[ES] + k_2[ES] = [ES](k_{-1} + k_2) \qquad \textbf{(c)}$$

At this point we take into account the conservation of mass, the total enzyme concentration $[E]_T$ is the sum of free enzyme and enzyme-substrate complex:

$$[E]_T = [E] + [ES] \quad \Rightarrow \quad [E] = [E]_T - [ES]$$

The result for [E] is substituted into eq. **c**:

$$k_1([E]_T - [ES])[S] = (k_{-1} + k_2)[ES] \qquad \text{(now we need to solve this for } [ES]\text{)}$$

$$k_1[E]_T[S] - k_1[ES][S] = k_{-1}[ES] + k_2[ES] \qquad \text{(brackets resolved)}$$

Box 3: continued

$$k_1[E]_T[S] = k_{-1}[ES] + k_2[ES] + k_1[ES][S] \quad (k_1[E][S] \text{ added to both sides})$$

$$k_1[E]_T[S] = [ES](k_{-1} + k_2 + k_1[S]) \quad ([ES] \text{ taken out as a factor})$$

$$[E]_T[S] = \frac{[ES](k_{-1} + k_2 + k_1[S])}{k_1} \quad (\text{both sides divided by } k_1)$$

$$\frac{[E]_T[S]k_1}{(k_{-1} + k_2 + k_1[S])} = [ES] \quad (\text{solve for } [ES])$$

$$\frac{[E]_T[S]k_1}{k_1\left(\frac{k_{-1}}{k_1} + \frac{k_2}{k_1} + [S]\right)} = [ES] \quad (\text{factor out } k_1)$$

$$\frac{[E]_T[S]}{\left(\frac{k_{-1} + k_2}{k_1} + [S]\right)} = [ES] \quad (k_1 \text{ cancels out})$$

$$\frac{[E]_T[S]}{K_M + [S]} = [ES] \quad \text{simplified with } K_M = \frac{k_{-1} + k_2}{k_1}$$

With this result (that is based on the steady-state assumption) we can go back into equation **a**:

$$v_0 = \left(\frac{d[P]}{dt}\right)_{t \approx 0} = \frac{k_2[E]_T[S]}{K_M + [S]}$$

This is the Michaelis-Mention equation 1a as shown in the text

3.3.1 Interpretation of the Michaelis-Menten equation

The Michaelis-Menten equation provides the dependence of the initial reaction rate v_0 on substrate concentration. The plot of v_0 against substrate concentration is called a *hyperbolic plot* or *saturation plot* and shown in Figure 17.

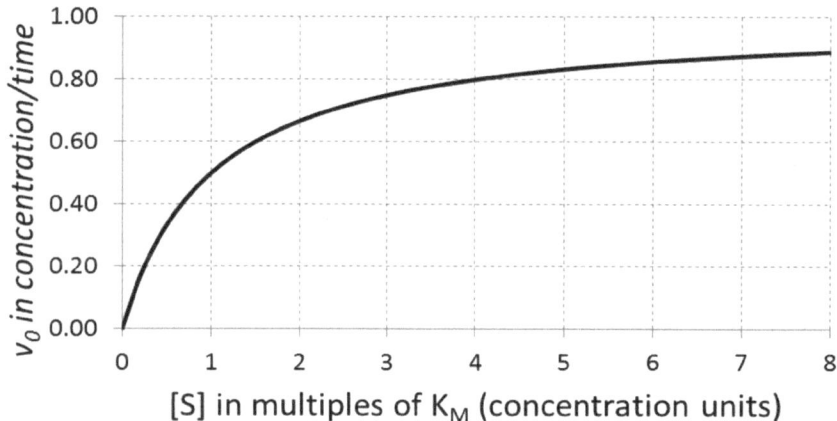

Figure 17: A plot of the initial reaction rate v_0 against substrate concentration according to the Michaelis-Menten equation with v_{max} = 1.0 concentration units/time. The substrate concentration is shown in multiples of K_M, so that K_M is equal to 1.0 in this example. Normally substrate concentrations would be in the units of µM or mM and v_0 in the units of µM/s or mM/s.

The typical feature of the saturation plot is the initial steep rise of v_0 with increasing substrate concentrations followed by a gradual flattening of the curve. The maximum initial reaction rate v_{max} is approached asymptotically. In the limit of infinite substrate concentration the v_{max} of 1.0 concentration/time is reached. Note that for the example shown in Figure 17 even at substrate concentrations of 8 K_M the maximum initial reaction rate has not been reached.

The aim of enzyme kinetic data analysis is to determine K_M and v_{max}. If we know the concentration of *active* enzyme sites, we can use v_{max} to calculate the *turnover number* that is the number of reaction the enzyme performs per unit time. This is also known as k_{cat}:

$k_{kat} = v_{max} / [E]_T$

Note that for multimeric enzymes, the concentration of active sites $[E]_T$ is a multiple of the enzyme concentration, while for enzymes with one active site $[E]_T$ is equal to the concentration of active enzyme. Furthermore, the concentration of active enzyme is often not equal to the total protein concentration, as proteins are sensitive biological materials that may be inactivated due to temperature, aggregation or heavy metal ions present as contaminants or contaminant microorganisms.

The Michaelis constant K_M corresponds to the substrate concentration at which the rate is half maximal, or in other words, the substrate concentration at which half the active sites of the enzyme are bound to substrate. K_M was defined as a composite of rate constants (see Box 6):

$$K_M = \frac{k_{-1} + k_2}{k_1}$$

The substrate binding reaction is often very fast, which means that the value of k_2 is often much smaller than k_{-1}, thus k_2 can be neglected compared to k_{-1}. In that case K_M can be approximated by k_{-1}/k_1, which is the equilibrium constant of the binding reaction (see Box 7) and related to the affinity of the enzyme for the substrate. Therefore, the Michaelis constant may be considered as an approximate measure of the *affinity* of the enzyme for its substrate. A *low* K_M means that the enzyme has a *high* affinity for its substrate. The ideal enzyme would have a low K_M and high k_{cat}, thus the expression

$$\frac{k_{cat}}{K_M} := \text{catalytic efficiency}$$

is defined as the catalytic efficiency. Some enzymes convert the substrate as fast as the substrate is able to reach the active site. Since the substrate in solution undergoes a process of diffusion to reach the active site the catalytic efficiency is limited by the speed of diffusion. Enzymes with a catalytic efficiency of 10^8 to 10^9 $M^{-1}s^{-1}$ have reached this *diffusion limit*.

Box 7: The relationship between kinetics and thermodynamics.

> In case of reversible reactions, the rate constants (kinetics) are related to the equilibrium constant (thermodynamics). For the simple example of
>
> $$A \underset{k_{-1}}{\overset{k_1}{\rightleftarrows}} B$$
>
> we can write the rate law for the forward reaction: $\quad v_{forward} = \dfrac{d[A]}{dt} = -k_{-1}[A]$
>
> and for the backward reaction: $\quad v_{backward} = \dfrac{d[B]}{dt} = -k_1[B]$
>
> And in equilibrium the forward and backward reaction rates are equal:
>
> $$v_{forward} = v_{backward}$$
> $$-k_{-1}[A] = -k_1[B]$$
> $$\frac{k_{-1}}{k_1} = \frac{[B]}{[A]} = K$$
>
> The last expression [B]/[A] is the equilibrium constant K for this reaction, a thermodynamic quantity.

3.3.2 Enzyme activity and enzyme units

When enzymes are sold the specific enzyme activity in enzyme *Units* per gram or per mL is typically specified. Enzyme preparations with a higher specific activity are sold at a higher price. The enzyme Unit (U) is defined as the amount of substrate in μmol transformed in one minute (60 s). Dependent on the stoichiometry of the reaction catalysed, it is often equal to the amount of product produced in one minute. The U is measured at 25°C, the optimal pH and the substrate concentration that gives the maximum reaction rate. The determination of U for an enzyme preparation is often based on a two point measurement, i.e. the change of substrate (or product) concentration within a defined time interval. This is an approximate measure as substrate concentrations that give the maximum reaction rate are difficult to achieve. A more accurate determination is possible based on a full kinetic analysis that determines the turnover number k_{cat}. The enzyme activity is then given by:

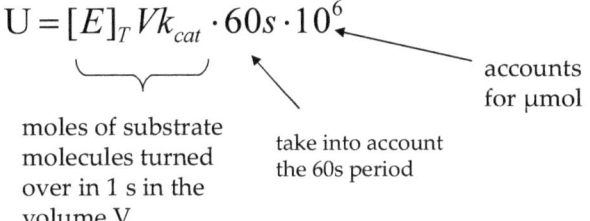

$$U = [E]_T V k_{cat} \cdot 60s \cdot 10^6$$

- $[E]_T V k_{cat}$: moles of substrate molecules turned over in 1 s in the volume V.
- $60s$: take into account the 60s period
- 10^6: accounts for µmol

Note that $[E]_T$ (in mol/L) refers to the concentration of enzyme active sites and not the total protein concentration.

The *specific activity* is the activity in U per mass (mg, g) of protein or per volume (mL, L) of solution. Often enzyme preparations are not pure and contain other proteins, thus with increasing purity the specific activity increases and the enzyme preparation can be sold at a higher price.

3.4 Analysis of enzyme kinetic data

The analysis of enzyme kinetic data is aimed at determination of v_{max} and K_M, and the potential influence of inhibitors on enzymes. The following steps are required for a full (steady-state) kinetic analysis of an enzyme (in the absence of inhibitors):

1) Measure how the product or substrate concentration 'c' changes over time.

2) Obtain the initial reaction rate $v_0 = \Delta c/\Delta t$ for the initial phase of the enzyme reaction (when the concentrations change in a linear fashion).

3) Repeat steps 1) and 2) for various substrate concentrations $[S]$ (ideally covering the range ½ K_M < $[S]$ < 5 K_M).

4) From your set of data $[S]_i$, $v_{0,i}$ obtain K_M and v_{max}.

With regards to point 3), it is important to choose substrate concentrations that cover the whole range of the hyperbolic initial rate curve (Figure 17). Since K_M is not know *a priori*, it may be required do to initial experiments covering a wide range of substrate concentration in order to estimate K_M, before choosing a number of substrate concentrations around K_M.

3.4.1 Obtain initial reaction rates

The primary data is usually a set of product or substrate concentrations obtained at different time points as shown in Figure 18.

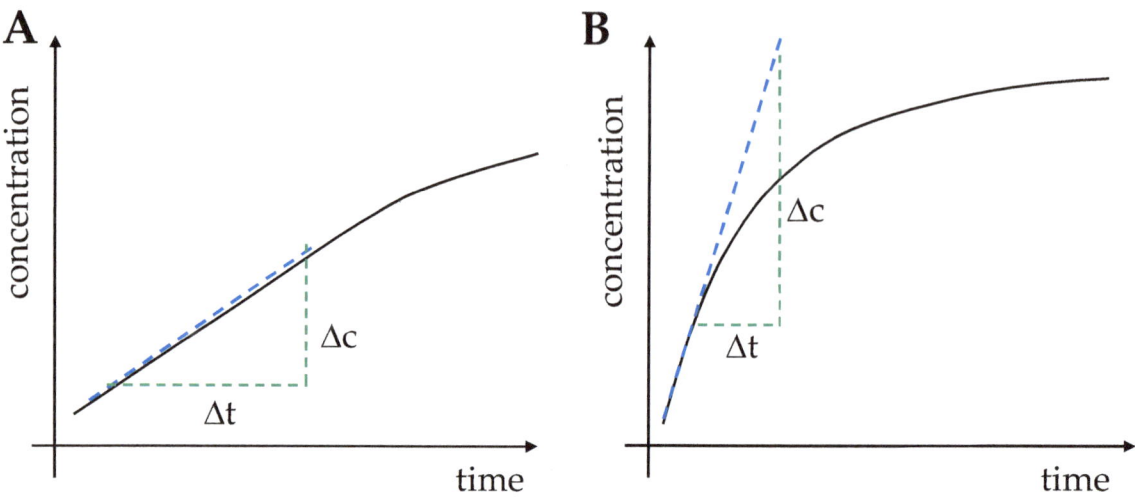

Figure 18: Change in product concentration over time as measured in an enzyme assay. The initial reaction rate is taken from the gradient of the linear phase as $\Delta c/\Delta t$. A) Example of a slow (compared to the timescale of the measurement) enzyme with a long linear phase B) Example of a fast enzyme with a short linear phase.

From the linear phase the initial reaction rate is determined as $v_0 = \Delta c/\Delta t$. If the enzyme is fast compared to the time scale of the measurement method as in Figure 18B, the determination of the initial rate becomes more inaccurate. Since at higher substrate concentration the reaction rate increases the time scale of the measurement may need to be changed (measuring at shorter time intervals).

For automated analysers used in biomedical laboratories often a two-point estimate of the reaction rate is taken. In the development of enzyme assays for automated analysers conditions need to be found that produce a linear concentration increase (or decrease) for a sufficient amount of time. The conditions that may be changed could be pH (different than optimal pH), temperature or a different substrate that gives lower reaction rates.

3.4.2 The hyperbolic plot

Once initial reaction rates have been obtained for different substrate concentrations, the Michaelis constant K_M and the maximum initial reaction rate v_{max} can be determined. Once possibility is to estimate these values from the hyperbolic (or saturation) plot as shown in Figure 19.

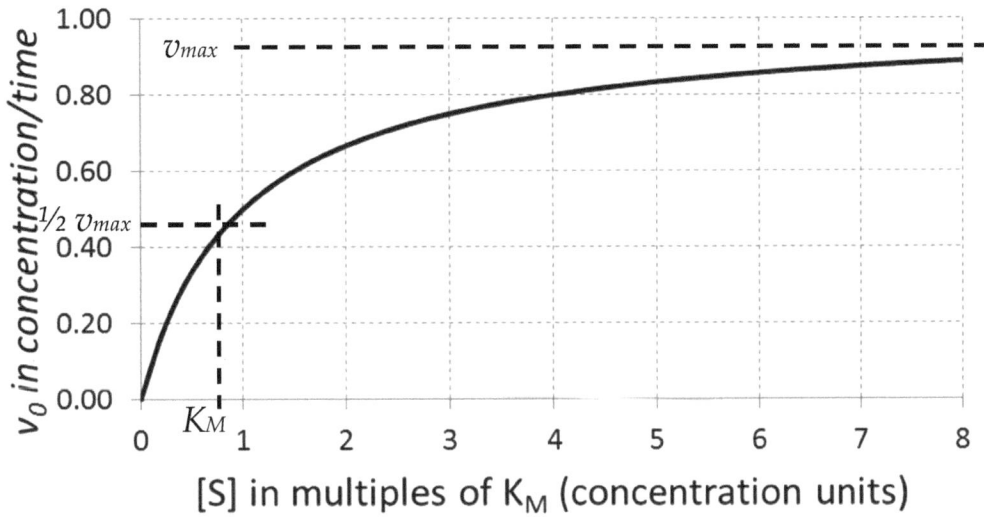

Figure 19: Illustration of the estimation of K_M and v_{max} from the hyperbolic plot. First v_{max} is estimated as 0.92 concentration units/time, then K_M is estimated as the substrate concentration that corresponds to ½ v_{max} as 0.90 concentration units. This curve was drawn with v_{max} = 1.0 concentration/time and K_M = 1.0 concentration units.

From Figure 19 it is apparent that the estimation of v_{max} is difficult since the curve approaches this value asymptotically. Even at a substrate concentration of eight times K_M the v_{max} estimate remains inaccurate. An inaccurate v_{max} leads to an inaccurate K_M, because K_M is determined at the substrate concentration that gives ½ v_{max}. Without widely available computers in previous times, scientists have a developed a variety of methods that transform the Michaelis-Menten equation into a linear form. The most popular of these linear transformations (commonly used in practicals for students) is the double reciprocal transformation by Lineweaver-Burk developed in 1934.

3.4.3 The Lineweaver-Burk plot

The basis of the Lineweaver-Burk plot (also known as double reciprocal plot) is taking the reciprocal of both sides of the Michaelis Menten equation (eq. 1b) leading to:

$$\frac{1}{v_0} = \left(\frac{K_M}{v_{max}}\right)\frac{1}{[S]} + \frac{1}{v_{max}} \qquad (2)$$

y = gradient·x + intercept$_y$

As it can be seen this expression has the form of a linear equation, if $1/v_0$ on the y-axis (ordinate) is plotted against $1/[S]$ on the x-axis (abscissa). See for Box 8 for further details. The gradient is equal to K_M/v_{max} and the intercept with y-axis is equal to $1/v_{max}$. It can be shown that the intercept with the x-axis is equal to $-1/K_M$.

Table 2: Data table for the Lineweaver-Burk plot shown in Figure 20

[S] / {concentration_units}	v_0 / {concentration_units time^{-1}}	1/[S] / {concentration_units^{-1}}	$1/v_0$ / {time concentration_units^{-1}}
0	0	-	-
0.20	0.17	5.00	6.00
0.40	0.29	2.50	3.50
0.60	0.38	1.67	2.67
0.80	0.44	1.25	2.25
1.00	0.50	1.00	2.00
2.00	0.67	0.50	1.50
3.00	0.75	0.33	1.33
4.00	0.80	0.25	1.25
5.00	0.83	0.20	1.20

Thus the reciprocal of both the substrate concentrations and the initial rates need to be taken as shown in Table 2. The Lineweaver-Burk plot of the data is shown in Figure 20.

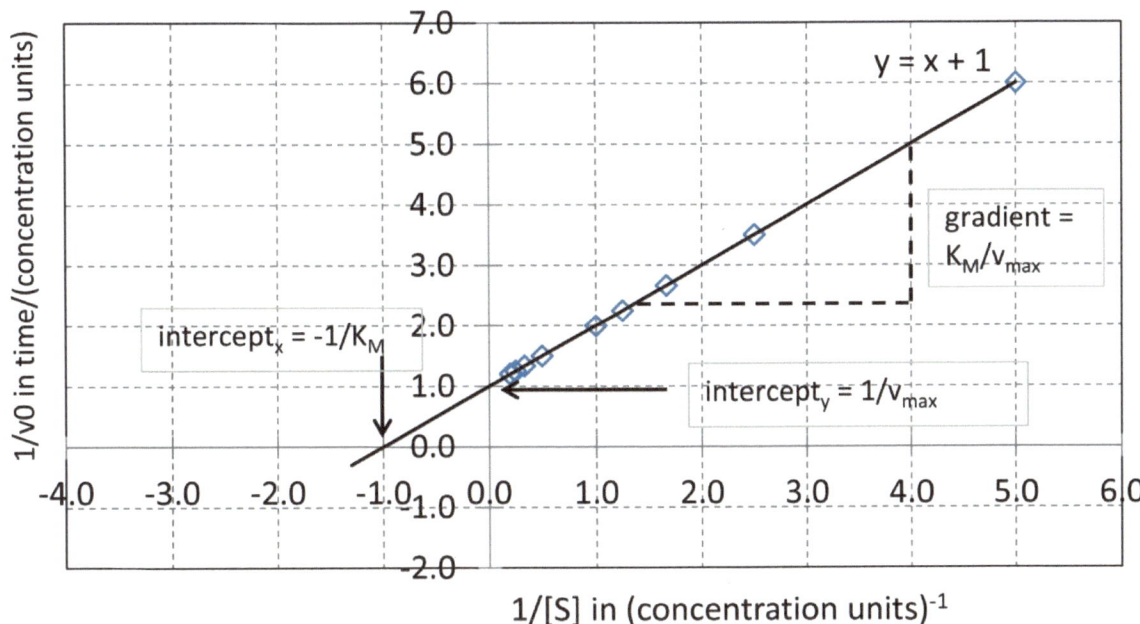

Figure 20: A Lineweaver-Burk plot of the same data as in Figure 19, with v_{max} = 1.0 (concentration units)/time and K_M = 1.0 concentration units. From the y-axis intercept v_{max} can be read out and from the x-axis intercept K_M can be read out. As the y-intercept is 1.0 time/(concentration_units), v_{max} is equal to 1/1.0 = 1.0 concentration_units time^{-1}. The gradient is 1.0 time, thus K_M = 1.0 time × 1.0 concentration_units time^{-1} = 1.0 concentration_units.

The enzyme kinetics parameters can be read out from the intercepts and/or the gradient as follows:

v_{max} = 1/intercept$_y$

K_M = gradient · v_{max}

Or: K_M = -1/intercept$_x$

The parameters can be read out more accurately than from the hyperbolic plot. A disadvantage of the Lineweaver-Burk plot is the unequal weighting given to the data points; most of the data points are compressed in the left part of the plot, while a high weight is given to a single data point at the lowest substrate concentration. Due to these disadvantages, non-linear curve fitting is the established procedure for initial rate data analysis in scientific research.

Box 8: Illustration of the concept and application of linear plots in science and reciprocal transformation of the Lineweaver-Burk equation.

Linear plots are very popular for data analysis in science as a straight line of best fit can be easily drawn with a ruler, while deviations from linearity can be clearly identified. Positive and negative gradients as well as intercepts can be found accurately.

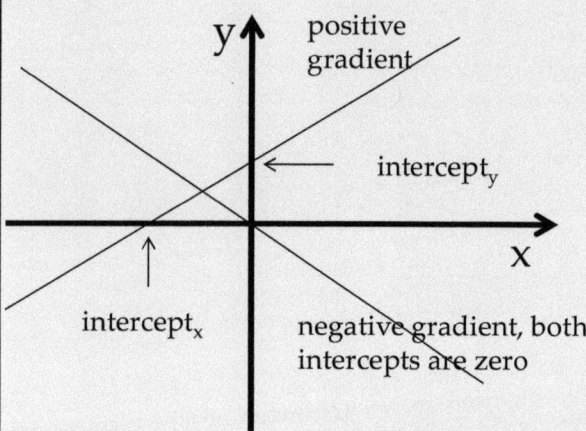

Figure showing two linear plots drawn into an x,y coordinate system.

The general equation for a straight line is:

y = gradient · x + intercept$_y$

The intercept with the x-axis is given as intercept$_x$ = -intercept$_y$ / gradient

Non-linear equations in science can sometimes be transformed into a linear form, the gradient and intercept then have a particular meaning. A few examples are given below:

p^2 = (a/b) t + 1/k

Plotting p^2 against *t* give a straight line with gradient *a/b* and y-intercept *1/k*.

ln A = -k t + ln A$_0$

Plotting *ln A* against *t* gives a straight line with gradient *–k* and y-intercept *ln A$_0$*.

Box 8: Continued

The reciprocal form of the Michaelis-Menten equation is obtained by taking the reciprocal of both sides:

$$(v_0)^{-1} = \left(\frac{v_{max}[S]}{K_M + [S]}\right)^{-1}$$

$\Leftrightarrow \quad \dfrac{1}{v_0} = \dfrac{K_M + [S]}{v_{max}[S]}$

$\Leftrightarrow \quad \dfrac{1}{v_0} = \dfrac{K_M}{v_{max}[S]} + \dfrac{[S]}{v_{max}[S]}$ ([S] cancels out)

$\Leftrightarrow \quad \dfrac{1}{v_0} = \dfrac{K_M}{v_{max}} \dfrac{1}{[S]} + \dfrac{1}{v_{max}}$

Thus a plot of $1/v_0$ (y-axis) against $1/[S]$ (x-axis) gives a straight line with gradient K_M/v_{max} and y-intercept $1/v_{max}$.

3.4.4 Non-linear curve fitting

Linear transformations are useful for data plotting and visual analysis but may also lead to an unequal weighting of data points and a distortion of errors. The established method in scientific research is non-linear curve fitting. This method relies on a computer-based algorithm that is aimed at minimising the deviation between the non-linear Michaelis-Menten curve and the experimental data points by varying K_M and v_{max} until the best fit between data points and curve is achieved. The agreement between the N data points (measurements of initial reaction rates at various substrate concentrations $[S]_i$) $v_1, v_2, \ldots v_i, \ldots, v_N$ and the values calculated from the Michaelis-Menten equation (equation 1b) $c_1, c_2, \ldots c_i, \ldots, c_N$ is judged based on the sum of the square deviations:

$$\text{square deviation} = \sum_{i=1}^{N}(c_i - v_i)^2, \text{ whereby the calculated values are } c_i = \frac{v_{max}^{estimate}[S_i]}{K_M^{estimate} + [S_i]}.$$

A computer algorithm tries to vary the estimates of K_M and v_{max} until the deviation is minimal. The method is also called 'least squares' curve fitting.

3.4.5 Fitting of differential equations

While the availability of computers to bioscientists has enabled the application of non-linear curve fitting, the increase in computer speed has enabled the fitting of systems of differential equations to whole progress curves (shown in Figure 5). As described in section 2.3 a reaction mechanism leads to a system of ordinary differential equations (ODEs) that can be integrated (or solved) numerically. With a given set of rate constants the ODEs can be solved and the resulting concentration-time data is compared to experimental data. The rate constants are varied until the minimum square deviation between experimental and calculated concentrations is reached. User interfaces have been developed that allow drawing the reaction scheme, while the system of ODEs is automatically derived and fitting to experimental data is performed (http://enzo.cmm.ki.si/) (Bevc et al, 2011) or Dynafit (Kuzmic, 2009) . These method do not require any approximations, such as the steady-state approximation, and measured concentrations over time can be used directly, rather than derived initial reaction rates. In fact, the linear phase of an enzyme reaction as shown in Figure 18 does not contain sufficient information, while using the whole progress curve (Figure 5) obtained at lower substrate concentration (compared to enzyme concentration) allows to obtain all rate constants (k_1, k_{-1} and k_2) of the Michaelis-Menten scheme from one experiment. Although that is not recommended, it illustrates the power of direct fitting of concentration-time curves to differential equations. Normally global fitting of several progress curves to a reaction meachanism would be performed.

3.5 Reversible inhibition

Enzyme inhibitors find application as medical drugs, as pesticides and herbicides in agriculture or as research tools to identify the role of enzymes in biological systems. For example penicillin is an inhibitor of the enzyme transpeptidase involved in bacterial cell-wall synthesis.

The binding of a *reversible* inhibitor to the enzyme can be reversed as opposed to an irreversible inhibitor. *Irreversible* inhibitors often have reactive functional groups that lead to a covalent modification of the enzyme. The extent of inhibition by irreversible inhibitors depends on the time of incubation. Inhibitor binding constants K_I or IC_{50} values (concentration that leads to 50% inhibition) are not defined. Some reversible inhibitors may bind slowly and with very high affinity, therefore they may look like irreversible inhibitors; those will not be considered. In the following we will cover the classical enzyme kinetic analysis of reversible inhibitors. Further, it is assumed that the enzyme-inhibitor complex is non-reactive; this type of inhibitor is called a *dead-end* inhibitor. Partial inhibitors are not considered here. Based on the Michaelis-Menten reaction scheme, reversible inhibitors can bind to the free enzyme's active site (competitive inhibitor), to the enzyme-substrate complex only (uncompetitive inhibitor) and to the free enzyme *and* the enzyme substrate complex (mixed inhibitor).

3.5.1 Competitive inhibition

In competitive inhibition the inhibitor molecule competes with the substrate for binding to the active site. The enzyme-inhibitor complex does not catalyse substrate – product conversion (dead-end inhibitor).

$$EI \underset{}{\overset{K_I}{\rightleftarrows}} I + E + S \underset{k_{-1}}{\overset{k_1}{\rightleftarrows}} ES \overset{k_2}{\rightarrow} P + E$$

There is a rapid equilibrium between the inhibitor I and the enzyme-inhibitor complex EI. The inhibitor dissociation constant is defined as: $K_I = [E][I] / [EI]$.

The inhibitor reduces the concentration of the free enzyme available for substrate binding. We, therefore, expect that a higher substrate concentration is required to achieve ½ v_{max}, i.e. the apparent K_M should increase with increasing inhibitor concentration, while v_{max} does not change. It can be shown that:

$$v_0 = \frac{v_{max}[S]}{\alpha K_M + [S]}; \quad \alpha = \left(1 + \frac{[I]}{K_I}\right) \qquad (3)$$

Equation 3 is the Michaelis-Menten equation for competitive inhibition. If we determine K_M in the presence of inhibitor, we would obtain an *apparent* K_M (also denoted as K_M'), which is not the true K_M of the enzyme, but a K_M modified by the influence of the inhibitor. The influence of the inhibitor concentration on K_M' is given by equation 3:

$$K_M' = \alpha K_M = \left(1 + \frac{[I]}{K_I}\right) K_M = K_M + \frac{K_M}{K_I}[I]$$

Note that rightmost expression has the form of a linear equation, thus by plotting K_M' against $[I]$, we obtain a straight line with y-intercept K_M and gradient K_M/K_I. This provides the means to obtain the inhibitor constant by determining the apparent K_M (K_M') at various inhibitor concentrations.

The mechanism of competitive inhibition can be visually identified from the Lineweaver-Burk plot as shown in Figure 21.

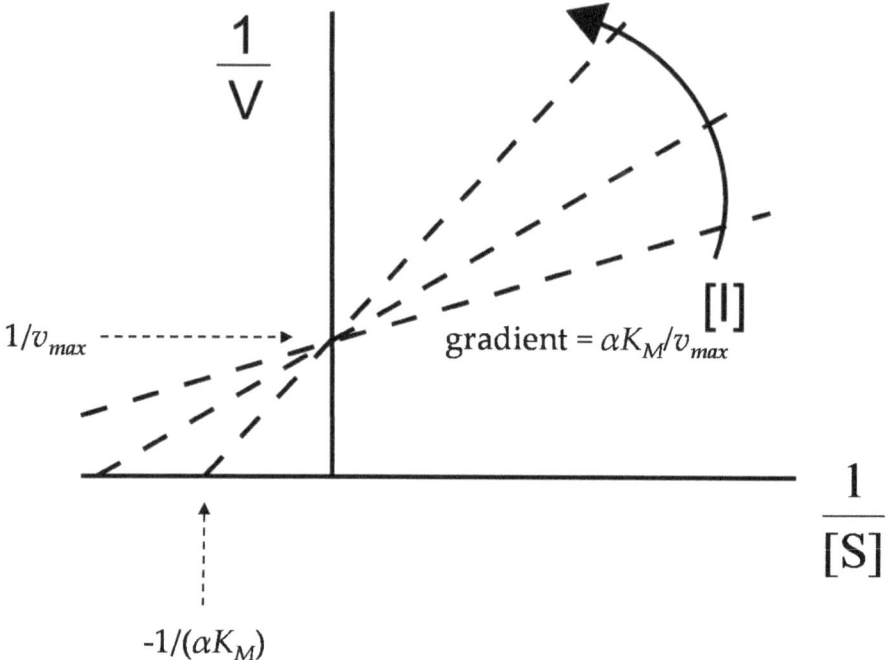

Figure 21: A Lineweaver-Burk plot for competitive inhibition. The data plotted at three different inhibitor concentration shows a set of lines that all intersect at the y-axis.

3.5.2 Uncompetitive inhibition

In this mode of inhibition the inhibitor binds only the enzyme-substrate complex:

$$E + S \underset{k_{-1}}{\overset{k_1}{\rightleftarrows}} ES \xrightarrow{k_2} P + E$$
$$+$$
$$I$$
$$\updownarrow K_I'$$
$$ESI$$

It can be shown that the Michaelis-Menten equation for uncompetitive inhibition changes to:

$$v_0 = \frac{v_{max}[S]}{K_M + \alpha'[S]}; \quad \alpha' = \left(1 + \frac{[I]}{K_I'}\right) \tag{4}$$

with $K_I' = [ES][I]/[ESI]$.

It can be shown from equation (4) that uncompetitive inhibition leads to a *decrease* of both the apparent v_{max} and the apparent K_M:

$$v_0 = \frac{v_{max}[S]}{K_M + \alpha'[S]} = \frac{v_{max}[S]}{\alpha'(1/\alpha' K_M + [S])} = \frac{v_{max}/\alpha'[S]}{1/\alpha' K_M + [S]}$$

Thus, the apparent v_{vmax} becomes $v_{max}' = v_{max}/\alpha'$ and the apparent K_M becomes $K_M' = K_M/\alpha'$. The Lineweaver-Burk plot in Figure 22 shows the characteristic picture of a series of parallel lines with increasing inhibitor concentration

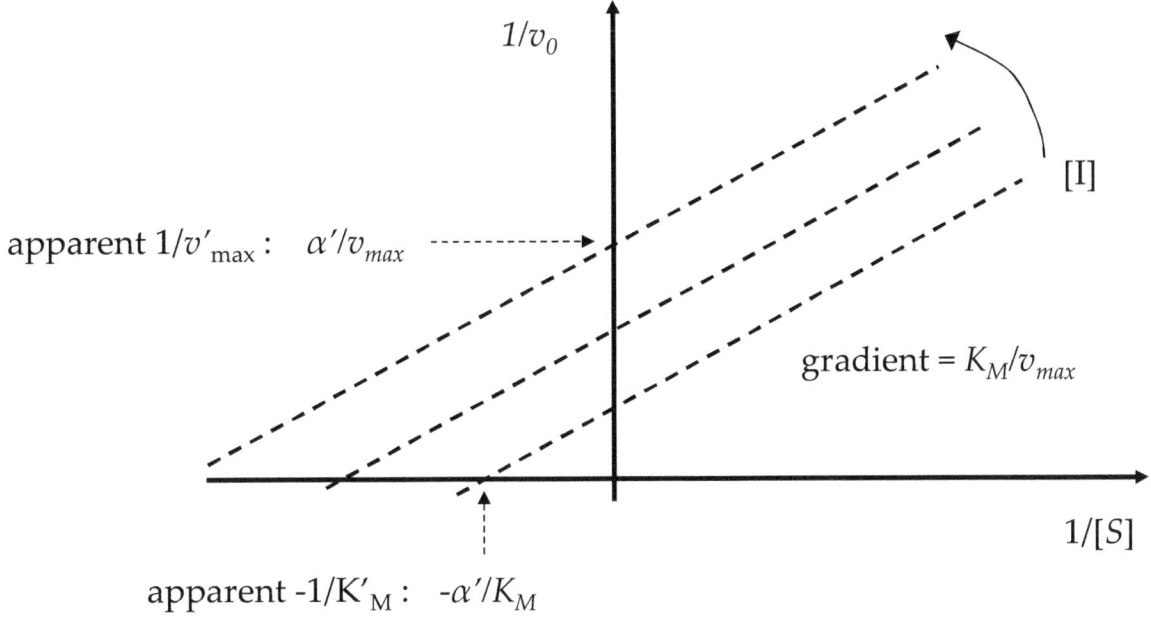

Figure 22: Lineweaver-Burk plot for uncompetitive inhibition with increasing inhibitor concentration [I].

3.5.3 Mixed and non-competitive inhibition

In mixed inhibition, the inhibitor binds both to the free enzyme and to the enzyme substrate complex:

$$EI \underset{}{\overset{K_I}{\rightleftarrows}} I + E + S \underset{k_{-1}}{\overset{k_1}{\rightleftarrows}} ES \overset{k_2}{\rightarrow} P + E$$

$$+ I$$

$$\updownarrow K'_I$$

$$ESI$$

The Michaelis-Menten equation for mixed inhibition changes to:

$$v_0 = \frac{v_{max}[S]}{\alpha K_M + \alpha'[S]}; \quad \alpha = \left(1 + \frac{[I]}{K_I}\right), \quad \alpha' = \left(1 + \frac{[I]}{K'_I}\right) \tag{5}$$

In the same way as for uncompetitive inhibition it can be shown that the apparent K_M becomes $K_M' = (\alpha/\alpha') K_M$ and the apparent v_{max} becomes $v_{max}' = v_{max}/\alpha'$. In mixed inhibition v_{max}' decreases with increasing inhibitor concentration, while K_M' may increase or decrease dependent on the α/α' ratio, and hence the K_I'/K_I ratio. The Lineweaver-Burk plot in Figure 23 shows an intersection point left of the y-axis.

A special case of mixed inhibition arises, if both binding constants K_I and K_I' are equal. This is called *non-competitive* inhibition and only v_{max} is influenced by the concentration of the inhibitor; the apparent v_{max} for non-competitive inhibition becomes $v_{max}' = v_{max}/\alpha$, so v_{max}' decreases with increasing inhibitor concentration. Although non-competitive inhibition is introduced here as a special case of mixed inhibition, it is found more often than mixed inhibition.

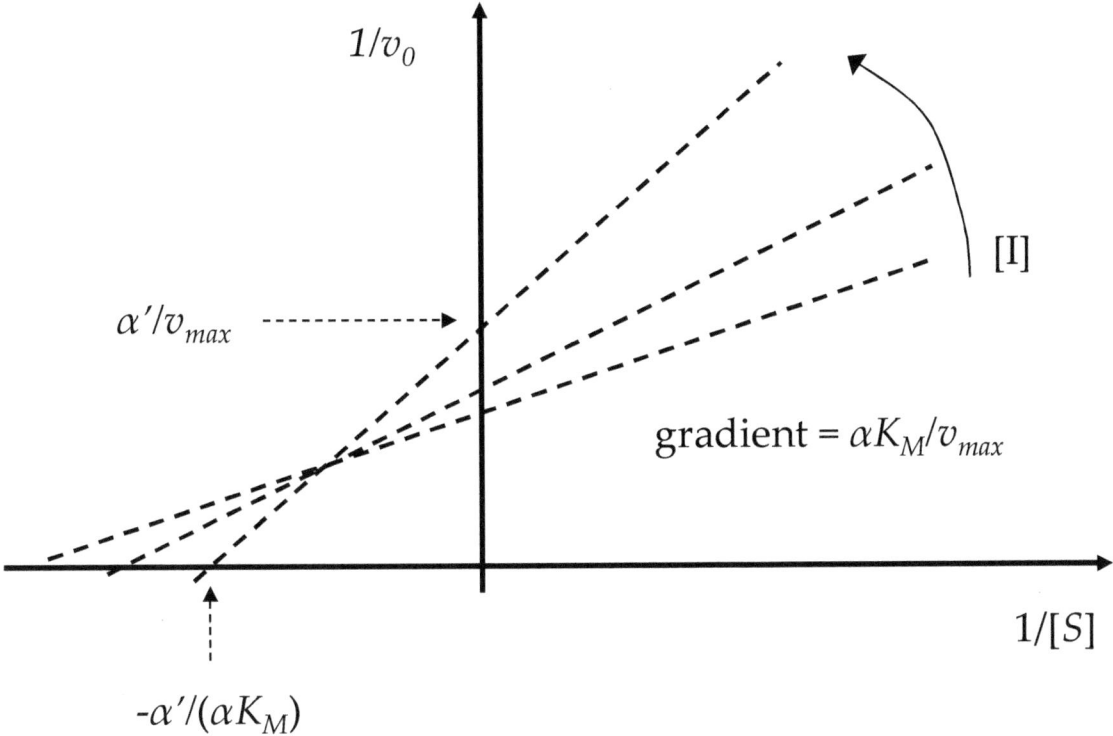

Figure 23: Lineweaver-Burk plot for mixed inhibition with increasing inhibitor concentration. With increasing similarity between the two dissociation constants K_I and K_I' the intersection point will move further left and eventually intersect at the x-axis, when $K_I = K_I'$. This is the special case of non-competitive inhibition.

3.5.4 Data analysis for reversible inhibition

The aim of the data analysis is to identify the mechanism of inhibition and then determine the numerical value of the inhibitor dissociation constant(s). Enzyme kinetics experiments must be performed with increasing concentration of enzyme inhibitor. Then the apparent v_{max} and K_M are determined and from the influence of the inhibitor on the apparent kinetic constants (Table 3) the mechanism of inhibition can be determined. It is very difficult to distinguish the mechanism of inhibition from the non-linear hyperbolic plot as illustrated in Figure 24.

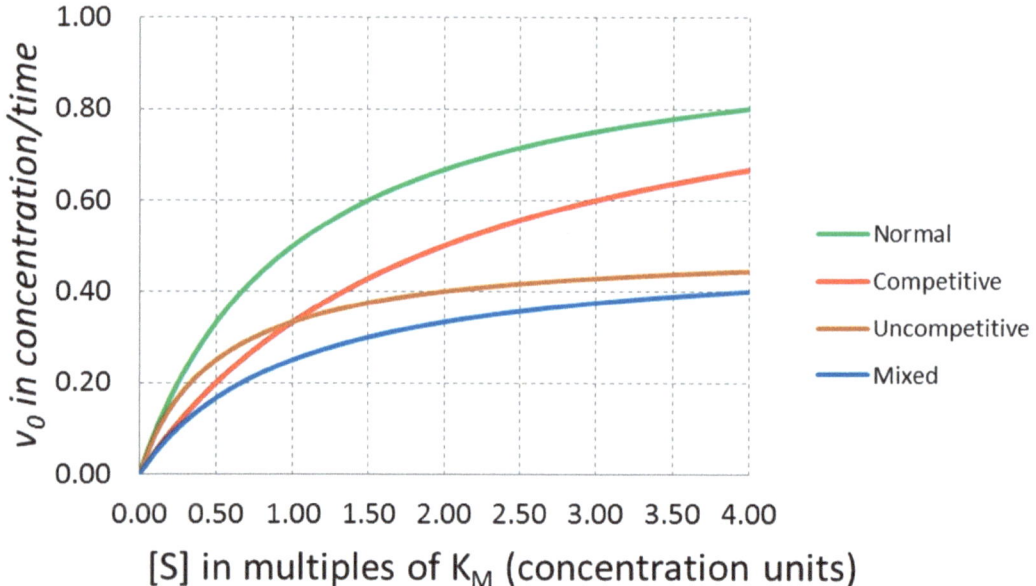

Figure 24: The hyperbolic plot of the initial reaction rate v_0 against substrate concentration [S] for an enzyme with v_{max} = 1.0 concentration/time and K_M = 1.0 concentration units under the influence of various types of reversible inhibitors.

A good diagnostic tool is the shape of the Lineweaver-Burk plots as shown in the sections above. At the same time the apparent kinetic constants should be determined and based on the influence of increasing inhibitor concentrations the mechanism of inhibition may be determined (Table 3).

Table 3: The change of the apparent enzyme kinetic constants with increasing inhibitor concentrations for different mechanisms of inhibition ('+' denotes increase, '−' denotes decrease and '0' denotes no change).

Mechanism	$K_M{'}$	$v_{max}{'}$
Competitive	+	0
un-competitive	−	−
Mixed	+/−	−
non-competitive	0	−

The next step is to perform a linear plot of the apparent kinetic constants against the inhibitor concentration as shown in Table 4.

Table 4: Determination of inhibitor dissociation constants from linear plots of apparent kinetic constants against inhibitor concentration [I]. Note that non-competitive inhibition is a special case of mixed inhibition with $K_I = K_I'$.

Mechanism	Linear Plot	Gradient	Intercept$_y$	Intercept$_x$
Competitive	K_M' against [I]	K_M/K_I	K_M	$-K_I$
Uncompetitive	$1/K_M'$ against [I]	$1/(K_M K_I')$	$1/K_M$	$-K_I$
mixed K_I'	$1/v_{max}'$ against [I]	$1/(K_I' v_{max})$	$1/v_{max}$	$-K_I'$
mixed K_I	K_M'/v_{max}' against [I]	$K_M/(v_{max} K_I)$	K_M/v_{max}	$-K_I$
non-competitive	$1/v_{max}'$ against [I]	$1/(K_I v_{max})$	$1/v_{max}$	$-K_I$

Alternatively, a combination of differential equations and binding equations may be fitted to inhibition data (Kuzmic, 2009)

3.5.5 Inhibitor binding sites for reversible inhibition

While the determination of the mechanism of inhibition does not allow to pinpoint the inhibitor binding site, it may lead to a hypothesis for the location of the inhibitor binding site relative to the enzyme active site. In that context we distinguish between isosteric (= same as active site) and allosteric (= elsewhere from the active site) binding. As illustrated in Figure 25 isosteric binding of the inhibitor can explain competitive inhibition; it is difficult to imagine how a competitive inhibitor could not bind to (or near) the active site. On the other hand the binding of the inhibitor to the free enzyme and the enzyme-substrate complex as it occurs in mixed inhibition hints at the possibility that the inhibitor may bind not at the active site but elsewhere on the enzyme (allosteric binding). Uncompetitive binding may be explained by the inhibitor binding near to the active site. However, the situation is not clear cut and a gradual transition between isosteric and allosteric binding can occur.

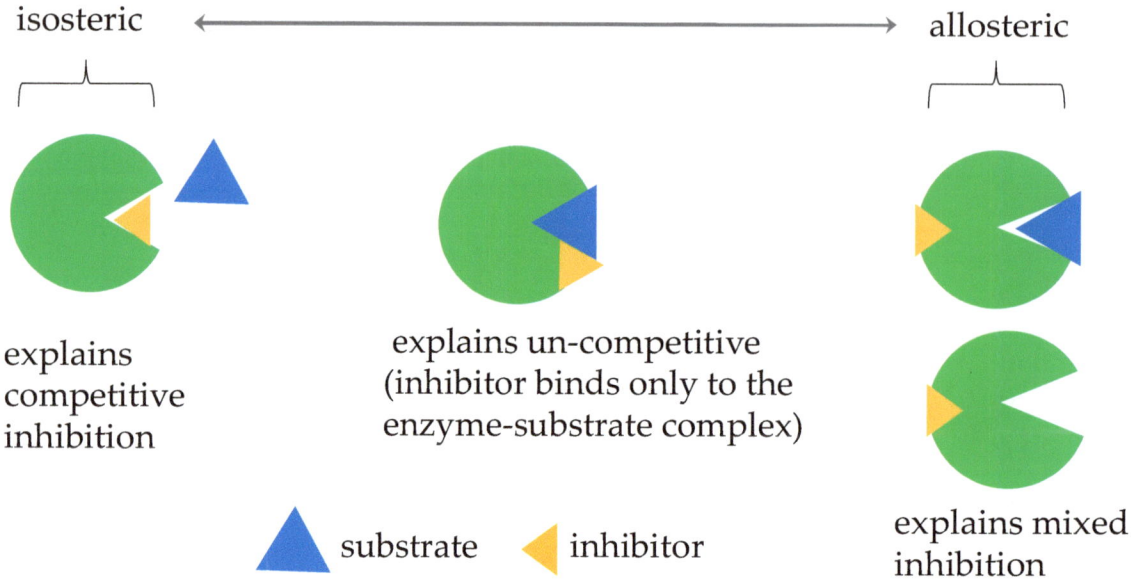

Figure 25: Hypotheses of inhibitor binding relative to the enzyme active site for various mechanisms of reversible inhibition.

3.5.6 Inhibitor constants in pharmacology

In classical enzyme kinetics an inhbitor is investigated by performing initial rate measurements at various substrate concentrations and by repeating those experiments at various inhibitor concentrations. From the results, the mechanism of inhibition and the inhibitor dissociation constant K_I can be inferred. The inhibitor dissociation constant K_I gives a complete thermodynamic description of the reversible binding equilibrium. In pharmacology it is often required to compare the effect of various inhibitors without too much experimental effort. For that purpose the enzyme activity is measured at one fixed (high) substrate concentration, while the inhibitor concentration is varied. As a result IC_{50} values are determined, that specify the concentration of inhibitor that yields 50% inhibition. An example of how an IC_{50} value is determined is shown in Figure 26.

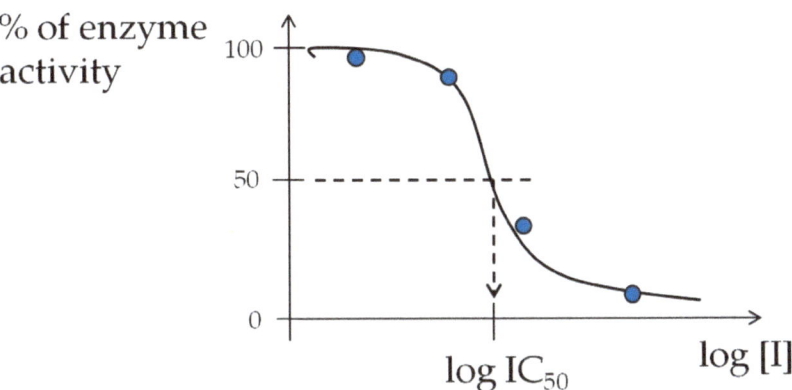

Figure 26: An example of the determination of an IC_{50} value for an inhibitor. The % of enzyme activity is plotted against the logarithm of the inhibitor concentration. The IC_{50} value can be determined graphically or better through non-linear curve fitting.

If the mechanism of inhibition, K_M of the enzyme, K_I of the inhibitor and the substrate concentration is known, the IC_{50} value can be calculated with the Cheng-Prusoff relationships, or alternatively K_I can be calculated from IC_{50} by rearringing the equations:

Competitive inhibition: $$IC_{50} = K_I\left(1 + \frac{[S]}{K_M}\right)$$

Uncompetitive inhibition: $$IC_{50} = K_I\left(1 + \frac{K_M}{[S]}\right)$$

Mixed inhibition: $$IC_{50} = \frac{[S] + K_M}{\frac{[S]}{K_I'} + \frac{K_M}{K_I}}$$

Non-competitive inhibition: $$IC_{50} = K_I$$

3 Enzyme Kinetics

Key points chapters 3-3.5

- Enzymes speed up chemical reactions. They do not change the equilibrium constant.
- Enzymes achieve this by stabilising the transition state.
- The majority of enzymes are protein molecules with a defined three-dimensional structure.
- The enzyme alkaline phosphatase has a complex mechanism that can be reduced to a simpler scheme.
- The Michaelis-Menten equation is based on the simplified mechanism

$$E + S \rightleftharpoons ES \longrightarrow E + P$$

- The steady-assumption is $d[ES]/dt = 0$ initially, which also means that the reaction rate at the beginning is constant, leads to the Michalis-Menten equation.
- The Michaelis-Menten equation is given as: $v_0 = \dfrac{v_{max}[S]}{K_M + [S]}$
- The Michaelis constant is K_M is the substrate concentration at ½ v_{max}.
- The turnover number k_{cat} is the number of reactions per active site per time ($v_{max} = k_{cat} [E]_T$).
- In an enzyme kinetics experiment the initial reaction rate is determined for a range of substrate concentrations.
- The Lineweaver-Burk plot (double reciprocal plot) is a linear plotting method for obtaining K_M and v_{max}.
- Non-linear curve fitting is a better method to obtain K_M and v_{max}.
- The Lineweaver-Burk plot is a useful tool to distinguish between the different types of reversible inhibition: competitive, uncompetitive and mixed (as well as non-competitive as a special case of mixed inhibition).
- An IC_{50} value can be converted into K_I (and vice versa) using the Cheng-Prusoff relations, if the inhibition mechanism, K_M and the substrate concentration is known.

3.6 Deviations from Michaelis-Menten kinetics

3.6.1 Substrate inhibition

Deviations from the hyperbolic curve describing the dependence of initial reaction rate on substrate concentration can be caused by substrate inhibition. In the first instance this may seem counterintuitive as more substrate would normally increase the reaction rate. There are normally multiple sites of interaction between the substrate and the enzyme. At very high substrate concentration, some of these interaction sites could be occupied by another substrate molecule, so that two substrate molecules are partially bound to the active site (Figure 27).

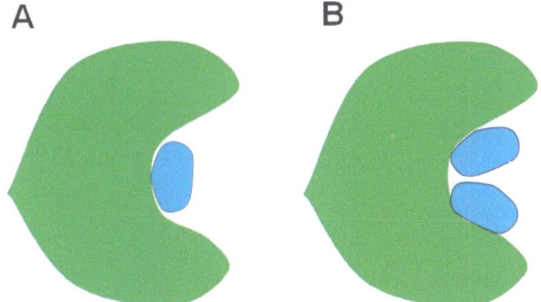

Figure 27: Cartoon depicting the binding of substrate (blue) to the enzyme active site A) at normal levels of substrate concentration, and B) at very high levels of substrate concentration leading to substrate inhibition.

It is conceivable that under such circumstances no reaction would occur. A variant of the Michaelis-Menten equation, which resembles the equation for uncompetitive inhibition, can describe the situation:

$$v_0 = \frac{v_{max}[S]}{K_M + [S]\left(1 + \frac{[S]}{K_i}\right)}$$

with K_i denoting the dissociation constant of the ES_2 complex. Substrate inhibition can be identified as a downwards deviation from the hyperbolic curve (Figure 28A) or as an upwards curve in the double reciprocal plot (Figure 28B). If the dissociation constant of the ES_2 complex is very large compared to K_M (unstable complex), the effect of substrate inhibition can be very subtle.

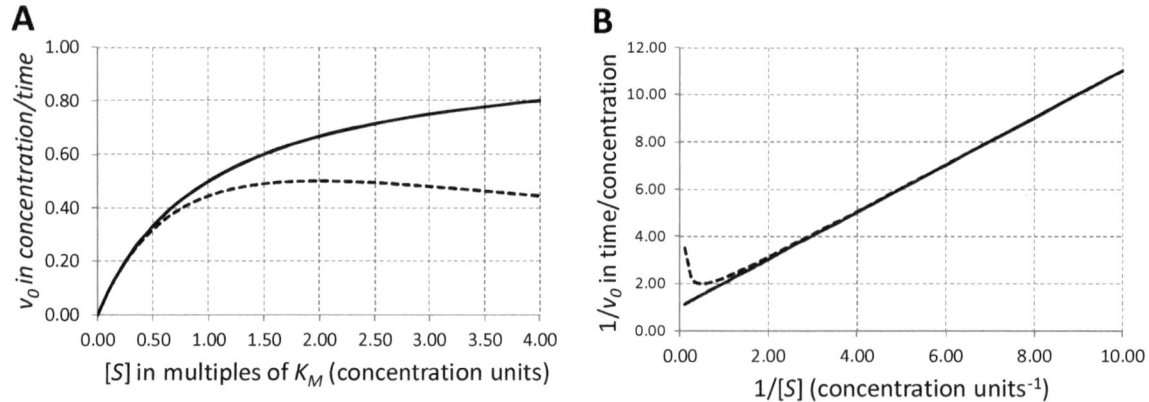

Figure 28: The case of substrate inhibition (dashed line) compared to the non-inhibited enzyme shown in A) the normal plot with linear axes and B) the double reciprocal plot. K_i was set to 4.0 in concentration units, K_M =1.0 and v_{max} = 1.0 concentration/time.

3.6.2 Product inhibition

Product inhibition is very common in enzyme catalysed reactions; as the products are released from the active site, they always have the capacity to bind to the active site. Enzymes in metabolic pathways are often controlled by product inhibition in order to ensure that only as much of a product is produced as is needed by the organism. In biotechnological processes, where enzymes are used as catalysts, product inhibition can be a problem, as the highest possible product concentration is desirable.

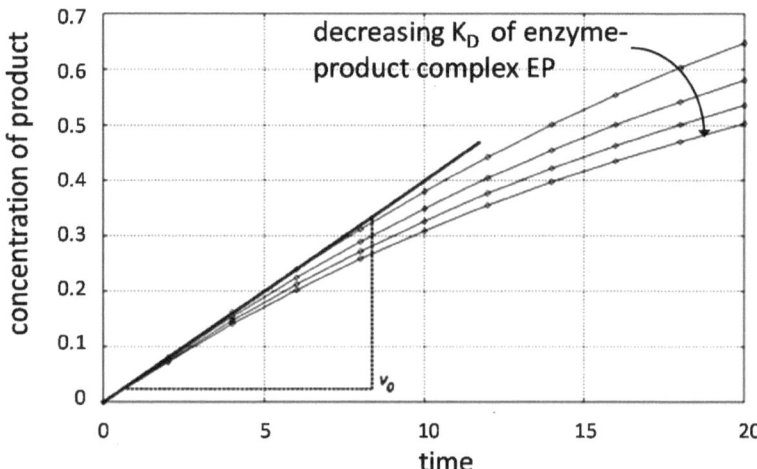

Figure 29: Example of product inhibition with decreasing dissociation constant K_D (increasing stability) of the enzyme-product complex. The concentration of product is plotted against time. The initial reaction rate v_0 is not affected.

Product inhibition is not detected in Michaelis-Menten type kinetic analyses of the initial reaction rate at various substrate concentrations. In the initial phase of the reaction the product concentration is small, so it has no measurable inhibitory effect (Figure 29). Product inhibition can be studied by adding product at the start of the reaction. The product often acts like a competitive inhibitor (3.5.1 Competitive inhibition).

3.6.3 Sigmoid kinetics and cooperativity

The Michaelis-Menten kinetics normally leads to a hyperbolic graph of the initial reaction rate against substrate concentration. Sometimes a sigmoid graph is observed as shown in Figure 30.

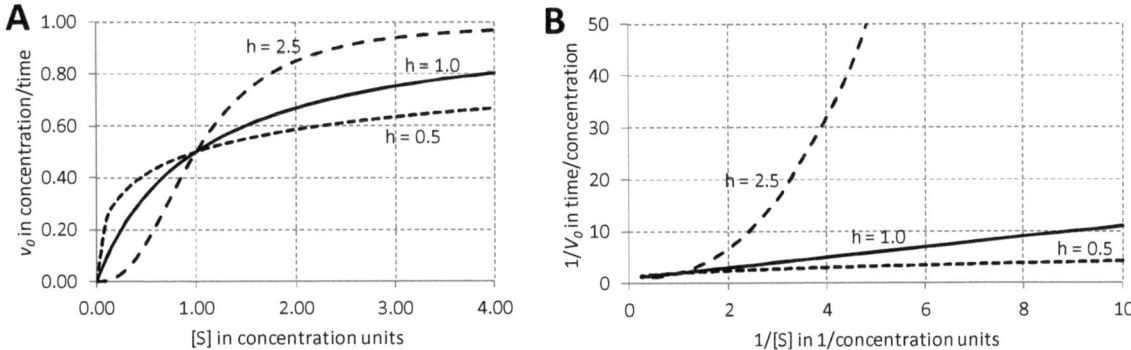

Figure 30: Saturation (A) and Lineweaver-Burk plots (B) of sigmoid curves with a Hill coefficient of h = 2.5 (positive cooperativity) and h = 0.5 (negative cooperativity). The solid curve with h = 1.0 illustrates the normal case (no cooperativity).

A sigmoid graph can be explained by *cooperativity* of substrate binding, in case the enzyme has more than one active site and binding to one site influences the binding to other sites. Binding sites could be on the same protein subunit, or more commonly one binding site on a subunit of a homo-multimer. *Positive cooperativity* occurs, if the binding to one site *increases* the affinity of other sites; the less common *negative cooperativity* occurs, if the binding to one site *reduces* the affinity of other sites. An equation describing cooperative protein-ligand binding has been developed empirically by Hill to describe the binding of oxygen to tetrameric haemoglobin. Applied to enzyme kinetics the equation is:

$$v_0 = \frac{v_{max}[S]^h}{K_{1/2}^h + [S]^h} \tag{6}$$

The constant $K_{1/2}$ is the substrate concentration at which $½ v_{max}$ is reached. The Hill-coefficient h is a measure of cooperativity, namely $h > 1$ for positive cooperativity and $h < 1$ for negative cooperativity. Although the Hill-equation was proposed empirically, simply to achieve a better fit to the data, it may be derived from a *hypothetical* binding reaction of substrate to an enzyme with n binding sites:

$$E + nS \rightleftarrows ES_n$$

This reaction describes an extreme case of cooperativity, in which all ligands bind in an all-or-none fashion. For this *hypothetical* reaction the Hill-coefficient would be $h = n$, the number of binding sites. In reality the Hill coefficient is $1 < h < n$, for positive cooperativity and $h < 1$ for negative cooperativity. Thus for positive cooperativity the next whole number following the Hill-coefficient is a lower boundary for the number of binding sites. For example, if $h = 2.5$, the enzyme would have *at least* three substrate binding sites.

Note that there are other mechanisms that may cause sigmoid kinetics, such as multiple enzyme conformations, while the enzyme has only one binding site:

$$E + S \rightleftarrows E^1S \rightleftarrows E^2S \rightleftarrows E^3S \text{ followed by product formation.}$$

Or:

$$E^1 \rightleftarrows E^2 \rightleftarrows E^3 \xrightleftharpoons{+S} E^3S \text{ followed by product formation.}$$

In the reaction schemes above $E^1, E^2, E^3, \ldots$ correspond to different enzyme conformations or substates. As there is only one binding site, sigmoid kinetics cannot be explained by cooperativity between binding sites. An example for an enzyme with one binding site that shows sigmoid kinetics is glucokinase (see

3.8 Allosteric enzymes). Glucokinase catalyses the first step in glycolysis, the formation glucose-6-phosphate. Glucokinease is one of the four hexokinase isoenzymes (also known as hexokinase IV) and is expressed in the liver and pancreas.

3.7 Enzyme reactions with two substrates

Most enzyme catalysed reactions involve two substrates, for example the hydrolysis of phosphate-esters (3.2 The reaction mechanism of alkaline phosphatase) involves the ester and water and the same is true for all other hydrolysis reactions. The phosphorylation by ATP-dependent kinase requires ATP in addition to the main substrate, for example hexokinase in glycolysis that catalyses the formation of glucose-6-phosphate. The majority of these reactions can be treated with Michaelis-Menten kinetics, by ensuring that the concentration of one of the substrates is high enough, so that it practically does not change during the measurement of the initial reaction rate. This is certainly true for water, while for the study of other reactions the experimenter normally chooses the conditions, such that the assumption of invariant concentration for one of the reaction partners is true.

Among the reactions with two substrates, the most common ones involve a ternary complex (E + A + B → EAB) or an enzyme intermediate that contains a group from one substrate associated with the enzyme (E + A-X → E-X + A). The latter type of mechanism is often called ping-pong or double displacement mechanism, since the group X is first displaced from the substrate and then displaced from the enzyme to form the final product. In both types of reactions there are *two* substrates and *two* products, thus it is called a *bi bi* reaction.

3.7.1 Ternary complex mechanisms

A + B ⇌ A' + B'   (A, B: substrates; A', B': products)

This mechanism can occur in two variants, the random order (order of substrate binding does not matter) and the compulsory order variant. The compulsory order mechanism involves the binding of two substrates and the release of the two products in a fixed order (A binding first in this example):

$$E + A \rightleftharpoons EA \xrightarrow{B} EAB \rightleftharpoons EA'B' \xrightarrow{B'} EA' \rightleftharpoons E + A'$$

The random order mechanism allows the binding of substrates in any order (see Box 9 for an alternative display of mechanisms with Cleland diagrams):

```
         B                    B'
E + A ⇌ EA ↘              ↗ EA' ⇌ E + A'
            EAB ⇌ EA'B'
E + B ⇌ EB ↗              ↘ EB' ⇌ E + B'
         A                    A'
```

When the concentration of one substrate is kept constant, while the concentration of the other substrate is varied, the initial reaction rate follows the Michaelis-Menten equation:

$$v_0 = \frac{v_{max}^{app}[A]}{K_M^{A,app}+[A]}$$ if the concentration of B is kept constant, and

$$v_0 = \frac{v_{max}^{app}[B]}{K_M^{B,app}+[B]}$$ if the concentration of A is kept constant.

The term 'app' in the equations above indicates that apparent constants are measured that depend on the fixed concentration of A or B, respectivly. The effect of *reducing* the concentration of B (and measuring v_0 at various [A]) is similar to mixed inhibition. This can be understood by considering that the ternary complex EAB *must* be formed for the reaction to proceed. Withdrawing B from EAB has the same effect as adding an inhibitor that binds to the enzyme substrate complex. B also binds to the free enzyme forming EB, thus withdrawing B from EB has the same effect as adding an inhibitor that binds to the free enzyme. An inhibitor that binds to the free enzyme and the enzyme-substrate complex is described by the mechanism of mixed inhibition. Thus two-substrate reactions that take place via a ternary complex (random or compulsory order) are readily identified with a Lineweaver-Burk plot (Figure 31). Note that in Figure 31 the effect of increasing [B] is *opposite* to the effect of increasing the concentration of a mixed inhibitor, as increasing [B] releases the 'inhibitory' effect.

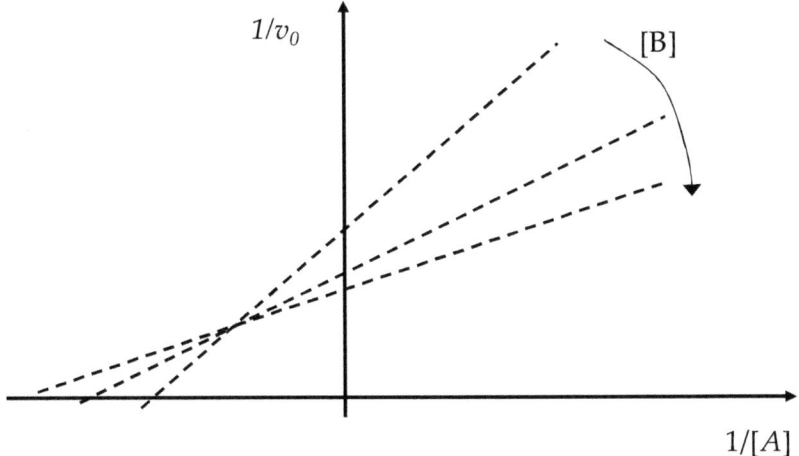

Figure 31: A Lineweaver-Burk plot for a two-substrate (A and B) enzyme reaction via a ternary complex. [B] is kept fixed at three different values, while the initial reaction rate is determined at various [A].

The y-intercepts $1/v_{max}^{app}$ in Figure 31 can be used to obtain v_{max} and K_M^B by a linear plot according to the equation:

$$\frac{1}{v_{max}^{app}} = \frac{1}{v_{max}} + \frac{K_M^B}{v_{max}}\frac{1}{[B]}$$

And the gradients $K_M^{A,app}/v_{max}^{app}$ can be used to obtain K_M^A by a linear plot:

$$\frac{K_M^{A,app}}{v_{max}^{app}} = \frac{K_M^B}{v_{max}} + \frac{K_M^A K_M^B}{v_{max}}[B]$$

It is not possible with these studies to distinguish between the random and compulsory order mechanism. This requires studies with competitive inhibitors for each substrate. For the example of a compulsory order mechanism (E → EA → EAB), a competitive inhibitor for B would lead to the formation of EAI. If the kinetics with inhibitor is measured through variation of the substrate [B], the *competitive* mechanism would be confirmed. If on the other hand, the kinetics were measured through the variation of the substrate [A], the inhibitor I would not compete with A, as EAI is formed. Thus the inhibitor would appear to bind only to the enzyme-substrate complex (EA), which is the mechanism of *uncompetitive* inhibition (see second line in Table 5).

In case of a random ordered mechanism (EA, EB, EAB are formed randomly) the competitive inhibitor for B would indeed show competitive inhibition, if [B] were varied as in the compulsory order case. If the kinetics were measured through the variation of [A], the inhibitor would not compete with A, as EA is formed. Thus the inhibitor would appear to bind to the free enzyme (EI) as well as to the enzyme substrate complex (EAI), which is the mechanism of mixed (or non-competitive) inhibition (see last line in Table 5). Through similar deductions the patterns of inhibition shown in Table 5 can be obtained.

As the inhibitor dissociation constants for the free enzyme and the enzyme substrate complex are very similar the special case of non-competitive rather than mixed inhibition is usually expected.

Table 5: Mechanism of inhibition observed for random and compulsory ordered two-substrate reactions using a 'competitive' dead-end inhibitor.

Mechanism	Inhibitor competes with	Inhibition mechanism observed	
		For varied [A]	For varied [B]
Compulsory ordered E → EA → EAB	A	competitive	non-competitive
Compulsory ordered E → EA → EAB	B	uncompetitive	competitive
Compulsory ordered E → EB → EBA	A	competitive	uncompetitive
Compulsory ordered E → EB → EBA	B	non-competitive	competitive
Random ordered	A	competitive	non-competitive
Random ordered	B	non-competitive	competitive

If no specific inhibitor can be found products can almost always be used as 'competitive' inhibitors. However, similar to the cases discussed above a 'competitive' product inhibitor can show different inhibition mechanisms dependent on the reaction mechanism, the order of substrate binding and in addition if one substrate is used at saturating or non-saturating concentrations. As an example consider a compulsory ordered mechanism with A binding first. The product inhibitor A' would lead to competitive inhibition for varied [A], if B were at saturating or non-saturating concentrations. For varied [B], the product inhibitor A' would not compete with B, but EA' (=EI, I: inhibitor) and EA'B (=EIS) complexes would form, which is characteristic of mixed (or non-competitive, if both binding constants are the same) inhibition. If a saturating concentration of the substrate A were used, there would be no inhibition, as A' would not be able to bind to the enzyme at all.

If for the same compulsory ordered mechanism (with A binding first) B' is used as a product inhibitor, for varied [A] there would be no competition but EB' (=EI) and EAB' (=ESI) would form, which is characteristic of mixed (non-competitive) inhibition. At saturated [B], EB and EAB would form and thus EA'B'. However, the next step, the release of B' (section 3.5.1) would be inhibited by addition of the product inhibitor B'. Since the release of B' from the enzyme is required before the release of A', the inhibition would take place at the level of the enzyme-substrate complex, which is a characteristic of uncompetitive inhibition. If B' would be studied as a product inhibitor for varied [B], the inhibition mechanism would be always non-competitive (mixed) at unsaturated and saturated [A]. Through similar logical reasoning the inhibition mechanism can be deduced for various combinations of product inhibitors, variation of substrate A or B at saturating or non-saturating concentrations of the other product. Please refer to more specialised literature for the complete pattern of product inhibition in random and compulsory ordered BI BI reactions (Copeland, 2000). In addition to logical reasoning illustrated above the inhibition pattern can also be deduced from the analysis of equations (Cornish-Bowden, 2012).

A variant of the compulsory ordered ternary complex mechanism is the Theorell-Chance mechanism, in which the second product B is transformed so quickly by the enzyme that the ternary complexes (EAB and EA'B') cannot be kinetically resolved.

$$E + A \rightleftharpoons EA + B \longrightarrow EA' + B' \rightleftharpoons E + A'$$

The Theorell-Chance meachanism can be confirmed with kinetic isotope studies, for example changing H-atoms against D-atoms (deuterium that has twice the atomic mass of hydrogen and normally slows down reaction rate). Another experimental approach is to change the viscosity of the solution at saturating concentration of the A substrate. If none of these measures affect k_{cat} or k_{cat}/K_M for substrate B, this can be taken as an indication of a Theorell-Chance mechanism.

3.7.2 Double displacement/ping-pong reaction

The double displacement or ping-pong reaction mechanism involves a transfer of a group from substrate A-X to the enzyme forming E-X. The group X may be covalently or non-covalently attached to the enzyme. Note that the reaction mechanism of alkaline phosphatase (chapter 3.2) is a ping-pong mechanism that involves the transfer of a phosphate group from para-nitrophenyl-phosphate to the Ser102 amino acid residue on the enzyme. This mechanism can be described as follows (see Box 9 for an alternative display):

$$E + AX \rightleftharpoons E{\cdot}AX \rightleftharpoons EX{\cdot}A \underset{A}{\rightleftharpoons} EX \underset{B}{\rightleftharpoons} EX{\cdot}B \rightleftharpoons E{\cdot}BX \rightleftharpoons E + BX$$

This is also a compulsory order mechanism; in case of a ping-pong mechanism substrate binding in random order would not make sense, as the chemical group X must be transferred to the enzyme first. In case of alkaline phosphatase the second substrate B is water, which is in abundance in aqueous solution, thus it is included in the rate constants and the mechanism can be analysed with the normal Michaelis-Menten equation. In general the double-displacement (ping-pong) mechanism can be detected by varying the concentration of the second substrate B. The substrate B only binds to the enzyme-substrate complex (EX·B), thus *reducing* the concentration of B has the same effect as *adding* an un-competitive inhibitor. The Lineweaver-Burk plot in Figure 32 shows the effect of increasing the concentration of the B substrate, which shows the opposite trend compared to adding an un-competitive inhibitor. The straight-line relationships in Figure 32 demonstrate that the Michaelis-Menten equation is applicable:

$$v_0 = \frac{v_{max}^{app}[AX]}{K_M^{AX,app} + [AX]} \quad \text{at constant } [B].$$

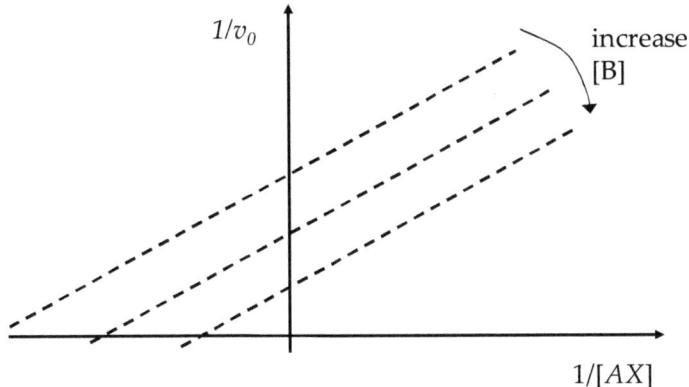

Figure 32: Lineweaver-Burk plot of two substrate enzyme reaction following the double displacement/ping-pong mechanism. The dependence of the initial reaction rate v_0 on the substrate concentration $[AX]$ was determined. The plot shows the effect of increasing the concentration of the second substrate B.

The apparent v_{max} and the apparent K_M obtained at various fixed B concentration can be used in further linear plots against $1/[B]$ to determine the true v_{max} and the true K_M^A and K_M^B.

$$\frac{1}{v_{max}^{app}} = \frac{1}{v_{max}} + \frac{1}{[B]}\frac{K_M^B}{v_{max}} \quad \text{and}$$

$$\frac{1}{K_M^{AX,app}} = \frac{1}{K_M^{AX}} + \frac{1}{[B]}\frac{K_M^B}{K_M^{AX}}$$

(y = intercept$_y$ + x gradient)

Box 9: An alternative display of two-substrate mechanisms using Cleland diagrams.

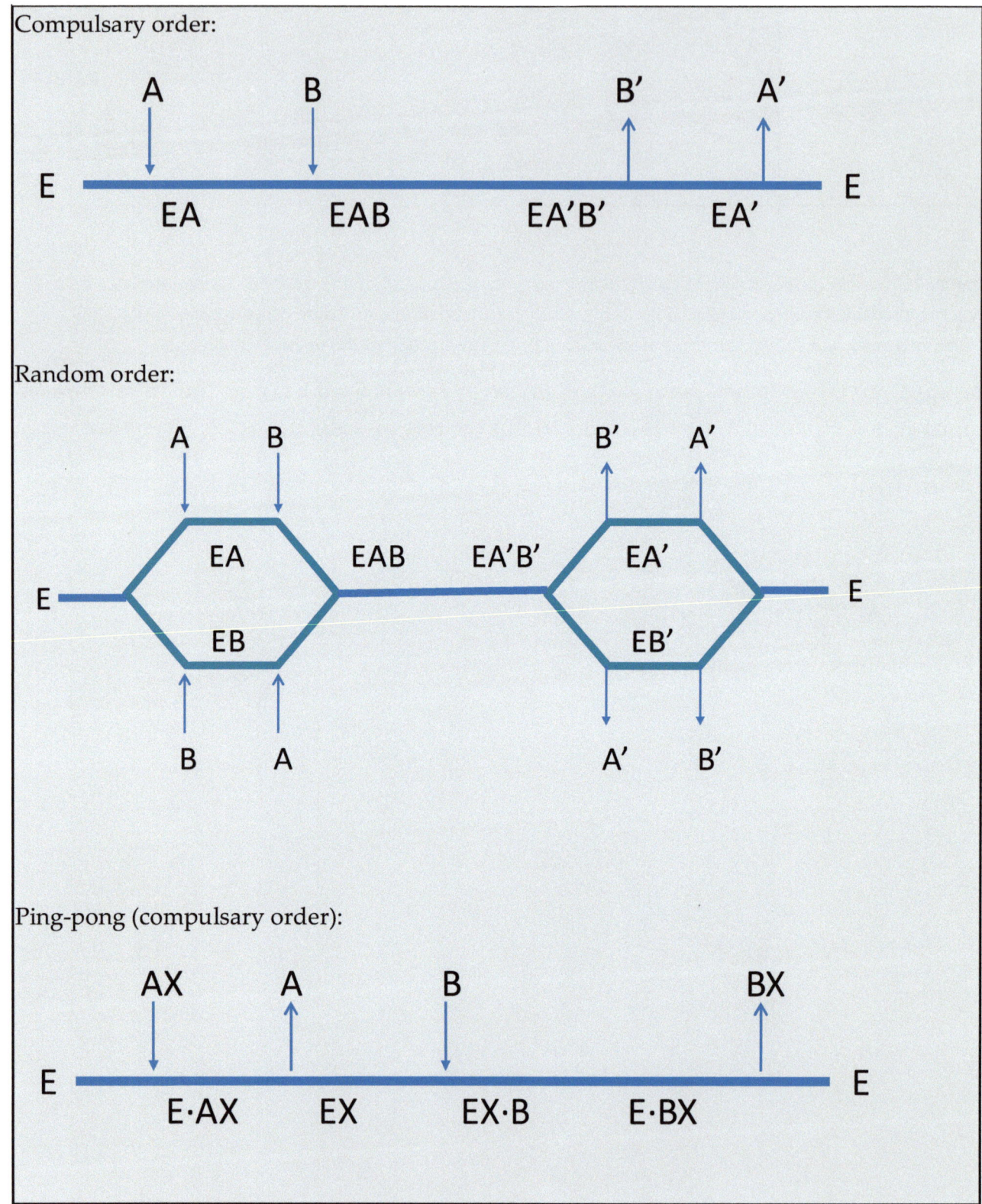

3.8 Allosteric enzymes

Allosteric enzymes have a binding site for an effector molecule that is *spatially distant* from the active site. This effector may be an activator or inhibitor. The binding site is called *allosteric site*; the effector is called *allosteric effector*. The allosteric effector is called *heterotropic*, if it is a different molecule than the substrate and *homotropic*, if it is the substrate. Indeed, the substrate can be an allosteric effector, if there is a binding site for the substrate that is spatially different from the active site of the enzyme.

An example of an allosteric enzyme is aspartate transcarbamoylase (ATC) that catalyses the first step in the synthesis of pyrimidine nucleotides. ACT is a multimeric protein that has twelve subunits; the catalytically active unit is a trimer C_3, and the regulatory unit is a dimer R_2, so the overal subunit composition is $2\,C_3 + 3\,R_2 = C_6 R_6$. Each catalytic trimer unit has three active sites and the regulatory dimer unit has two allosteric sites. The allosteric sites can bind the nucleotide ATP, an allosteric activator, and the nucleotide CTP, an allosteric inhibitor.

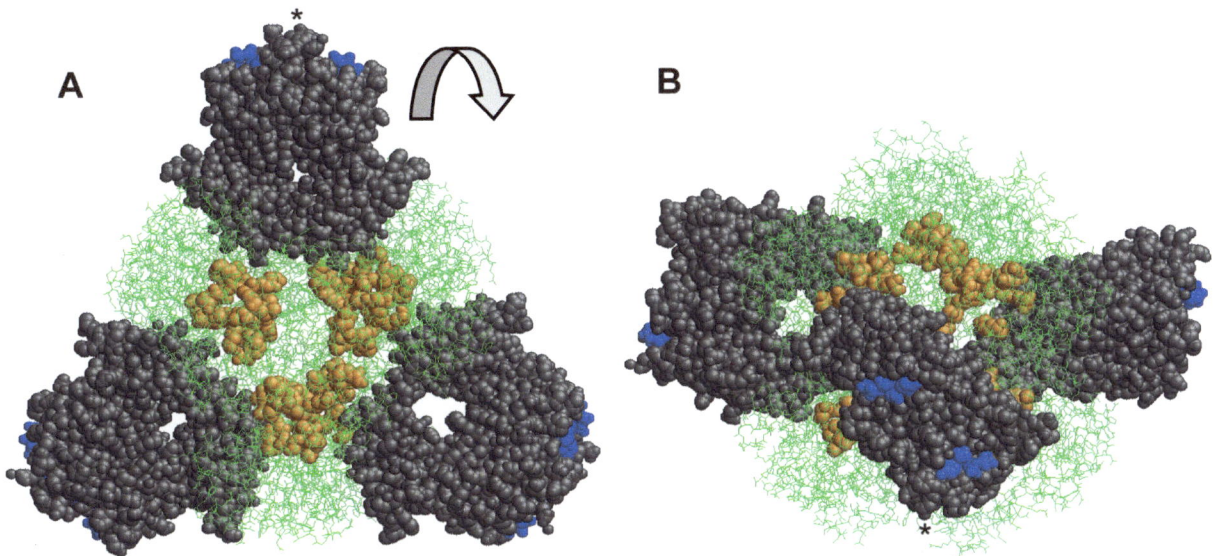

Figure 33: Structure of aspartate transcarbamoylase (PDB-ID: 4FYW). The two catalytic trimers are shown in green with active sites highlighted in gold. The three regulatory dimers are shown in grey with bound CTP in blue. B shows the structure at a different angle, rotated 90° to the front.

From the structure shown in Figure 33 it is clear that the regulatory sites (blue) are spatially distinct from the active sites (gold), hence ATC is an allosteric enzyme. Most allosteric enzymes (but not all) show cooperativity of substrate binding that can be identified from sigmoid initial rate kinetics (see page 60). An allosteric activator such as ATP shifts the sigmoid curve upwards and an allosteric inhibitor shifts the sigmoid curve downwards as shown in Figure 34.

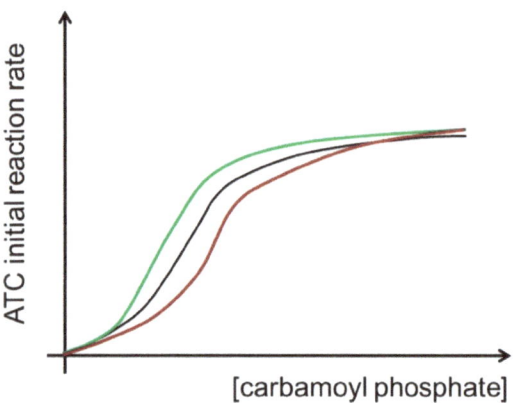

Figure 34: Initial reaction rate of aspartate transcarbamoylase (ATC) against substrate carbamoyl phosphate concentration. The sigmoid shape of the curves is characteristic of allosteric enzymes. Green curve: with addition of ATP, red curve: with addition of CTP, black curve: no addition.

Historically, the concept of allostery has been closely linked to cooperativity. Two models of substrate binding were developed for multi-subunit proteins, namely the symmetry (or concerted) model proposed in 1965 by Monod, Wyman and Changeux, and the sequential model proposed 1966 by Koshland, Némethy and Filmer. Both models assume that each subunit can exist in a tense and relaxed state, with the relaxed state having a higher affinity for binding of the substrate.

The symmetry model (Figure 35A), initial derived for oxygen binding to haemoglobin, assumes that subunits are symmetrically arranged. The substrate can bind to either the tense or relaxed state. The conformational change from the tense (square) to the relaxed state (circle) happens in a concerted way for all subunits. Initially, it was proposed that this conformational change affects the quaternary structure, i.e. a change in the orientation of subunits. According to structural studies of ACT it is now known that quaternary as well as tertiary structure is affected. The substrate binds with higher affinity to the relaxed state, thus substrate binding promotes the conformational change to the relaxed states and subsequently the second ligand will bind more strongly. This explains positive cooperativity. Negative cooperativity cannot be explained with the symmetry model. An allosteric regulator would stabilise the relaxed states (allosteric activator), or stabilise the tense state (allosteric inhibitor). The model makes a clear distinction between substrate binding and conformational states of the protein.

The sequential model combines substrate binding and conformational change (Figure 35B). Binding of the substrate to the tense state causes a change to the relaxed state, a kind of 'induced fit' process . The relaxed conformation of one subunit influences the conformation of the other subunit slightly (into a pre-relaxed state) increasing the affinity for the substrate, thus explaining positive cooperativity. In the sequential model the relaxed state is always substrate bound, we do not have any 'free' relaxed state as in the symmetry model.

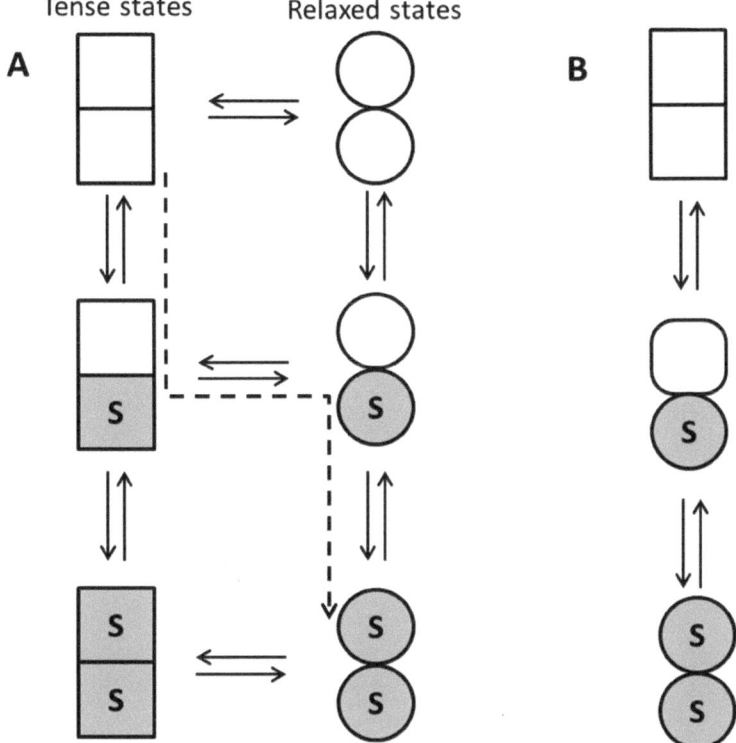

Figure 35: Models of allosterism for a dimeric protein, a square symbolises a tense conformation and a circle symbolises a relaxed conformation. A: The Monod, Wyman and Changeux symmetry/concerted model of allosterism. The dominant pathway is indicated with the dashed line. B: The Koshland, Némethy and Filmer sequential model of allosterims.

The sequential model can also address negative cooperativity, if the binding of the first substrate decreases the affinity of the other subunit for the substrate.
An allosteric inhibitor in the sequential model would stabilise the tense state, while an allosteric activator would stabilise the pre-relaxed state.

A mathematical analysis of both models can explain the sigmoidal shape of the initial rate curves of allosteric enzymes, such as aspartate transcabamoylase (with three substrate binding sites), thus both models are equally valid. However, the symmetry model is used more often in the literature. Although, these intitial models of allostery link cooperativity with allosteric control, later it was realised that also monomeric enzymes with only one active site showed sigmoid initial rate kinetics and are subject to allosteric control. Enzymes with one active site cannot, by definition, have cooperation between different acitve sites. The cooperativity in this case is between substrate binding and conformational change that explains the sigmoid kinetics. Thus, monomeric enzymes that show a high degree of structural flexibility may show sigmoid kinetics and allosteric control. An example is mammalian glucokinase, also known as hexokinase IV, that catalyses the formation of glucose-6-phosphate in the liver and pancreas. Glucokinase shows sigmoidal initial rate kinetics and an allosteric binding site for synthetic activators (Figure 36). In the case of glucokinase the sigmoidal initial rate kinetics is modelled by assuming at least two different protein conformations (high affinity and low affinity) that are able to bind glucose. This is supported by various structural studies that show that glucokinase is a highly flexible enzyme that can adopt various three dimensional structures.

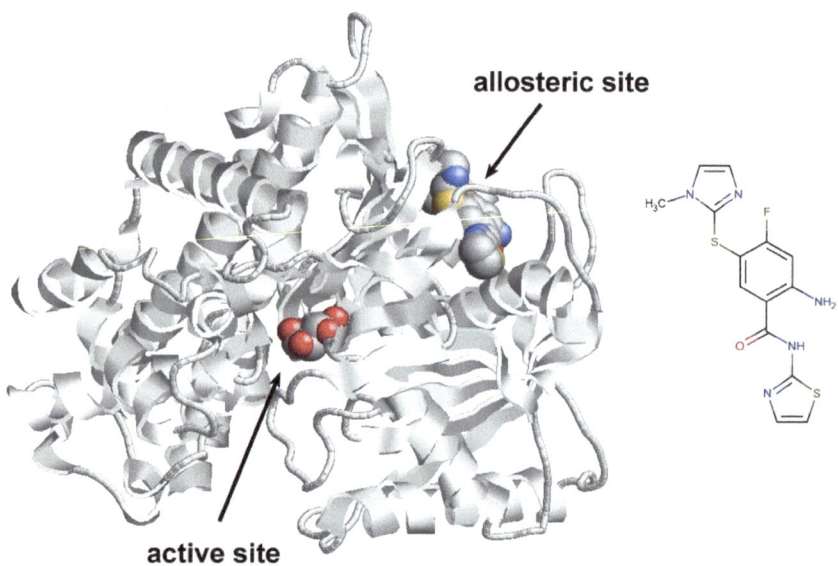

Figure 36: Ribbon diagramm of the three-dimensional structure of hexokinase showing glucose bound to the active site and the allosteric activator 2-amino-4-fluoro-5-[(1-methyl-1H-imidazol-2-yl)sulfanyl]-N-(1,3-thiazol-2-yl)benzamide (right) bound to the allosteric site (this figure was prepared from PDB-ID 1V4S).

A current view of allostery takes into account an ensemble of conformations that may range from just two distinct conformations for structurally stable proteins to a multitude of fast interconverting conformations for intrinsically disordered proteins. The study of the mechanisms of allostery is an active research area in structral biology and beyond the scope of 'Enzyme Kinetics Lecture Notes'.

Key points for chapter 3.6 Deviations from Michaelis-Menten kinetics and 3.7 Enzyme reactions with two substrates

- Substrate inhibition can be detected as upward curvature close to the y-axis in Lineweaver-Burk plots.

- Product inhibition is very common, but does not influence initial rate measurements.

- Sigmoid kinetics of initial reaction rates may occur due to positive cooperativity. Negative cooperativity does not lead to sigmoid curves.

- The most common two substrate reactions are double displacement/ping-pong and ternary complex mechanisms (random or compulsory order).

- Lineweaver-Burk plots of initial rate kinetics can distinguish between the two mechanisms.

- For the ternary complex mechanism, the order can be established with inhibitor studies.

- Allosteric enzymes have a regulatory binding site spatially distinct from the active site.

- Allosteric enzymes show often sigmoid kinetics, even if they have only one substrate binding site.

Reading for chapter 3:

Bevc S, Konc J, Stojan J et al (2011) ENZO: a web tool for derivation and evaluation of kinetic models of enzyme catalyzed reactions, *PLoS ONE* **6**: e22265

Copeland RA (2000) *Enzymes*, 2nd edition edn. Weinheim: Wiley-VCH.

Cornish-Bowden A (2012) *Fundamentals of enzyme kinetics*, 4th edn. Weinheim: Wiley-VCH.

Holtz KM, Kantrowitz ER (1999) The mechanism of the alkaline phosphatase reaction: insights from NMR, crystallography and site-specific mutagenesis, *FEBS Lett* **462**: 7-11

Kuzmic P (2009) Dynafit-a Software Package for Enzymology, *Method Enzymol* **467**: 247-280

4 SINGLE-MOLECULE KINETICS

The observation of single bio-molecules goes back to 1961, when Boris Rothman using fluorescence observed the activity of single β-D-galactosidase enzymes, that were highly diluted and dispersed in a 'water-in-oil' emulsion. Using the substrate o-nitrophenol-galactose that yields increased fluorescence when hydrolysed, discrete levels of fluorescence intensity were observed that corresponded to zero (background), one and two enzymes in a 'water-in-oil' droplet. In 1970 Hladky and Haydon observed the ion conductance of single ion channel molecules based on the measurement of electric current through a lipid bilayer formed across a tiny hole separating two aqueous compartments. Using a similar measurement principle with the lipid membrane formed at the tip of a glass pipette, Neher and Sakmann developed the well-known patch-clamp technique in 1981. The majority of single molecule measurements were based on ion conduction through ion channel proteins until in 1990 the Keller group showed the detection of single fluorescent molecules in solution. Since that time the field of studying single biomolecules has widened considerably, because fluorescent tags can be attached to a variety of biomolecules such as lipids, proteins, DNA and RNA. And with Fluorescence Resonance Energy Transfer (FRET) the change of distance between two fluorescent tags can be observed. Following on the from studies of Boris Rothman in 1961, the action of enzymes has been studied frequently by single-molecule fluorescence techniques. The study of single molecules in biology is important, as the number of molecules in intracellular compartments can be very small up to the limit of one DNA molecule per cell (Makarov, 2015).

4.1 Single molecule chemical kinetics

The principles of chemical kinetics apply to single molecules as well as to ensembles of molecules (ensemble kinetics). In fact, the ensemble kinetics is a special case of single molecule kinetics that is valid for large numbers of molecules. Single molecule kinetics is thus a more fundamental theory based on statistical physics. Note that in the following single molecule kinetics is compared to ensemble kinetics for educational purposes as ensemble kinetics was introduced first in this book; in a strict sense ensemble kinetics should be derived from single molecule kinetics.

4 Single-Molecule Kinetics

4.1.1 Systems of unimolecular reactions

Consider the following unimolecular elementary reaction:

$$S_1 \xrightarrow{k_{12}} S_2$$

The ensemble reaction rate is given as $\frac{d[S_1]}{dt} = -k_{12}[S_1]$ or in another form $d[S_1] = -k_{12}[S_1]dt$, which means that in a given time interval dt, the concentration of S_1 will change by $-k_{12}[S_1]dt$. For a few molecules S_1 or even a single molecule it does not make sense to specify concentrations; instead we are using the numbers of molecules directly:

$$\frac{dn_{S1}}{dt} = -k_{12}n_{S1} \text{ or in another form } dn_{S1} = -k_{12}n_{S1}dt$$

The rate constant specifies a probability per time that a reaction occurs or in other words $k_{12}dt$ is the probability that a reaction occurs in the time interval dt.

Instead of using numbers of molecules, we may use probabilities, e.g. $p_{S1} = n_{S1}/n_{S1(total)}$. Thus the differential equation can be rewritten as:

$$\frac{dp_{S1}}{dt} = -k_{12}p_{S1} \tag{7}$$

In case of a first order reaction the rate constant and kinetics of the single molecule situation is *identical* to the ensemble situation. In single molecule first order kinetics the term 'transition probability' is often used instead of 'rate constant'. An important assumption underlying single molecule (and ensemble kinetics) is that the probability for a molecular reaction to occur does not depend on the history of the molecule. This kind of process is known as a *Markov process*.

A particular aspect of single molecule kinetics is the possibility of observing reactions at equilibrium or steady state:

$$S_1^* \underset{k_{21}}{\overset{k_{12}}{\rightleftarrows}} S_2$$

If S_1 were fluorescent (as indicated by the '*' symbol) and S_2 not fluorescent or of lower fluorescence intensity, we would be able to see the switching between S_1 and S_2 as shown in Figure 37. The data shown is called a single molecule trace. The time a molecule spends in a certain state is variable and related to the kinetics of system described by the set of elementary reactions

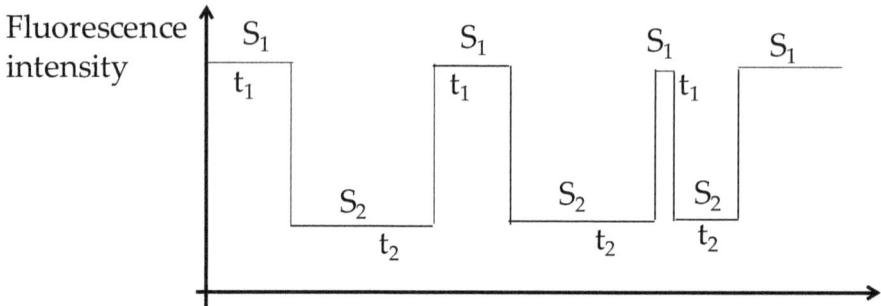

Figure 37: A idealised single molecule experiment showing equilibrium fluctuations between the high fluorescent S₁ state and the low fluorescent S₂ state. In reality the fluctuations would be overlaid by experimental noise. The succession of times the molecule stays in S₁ are denoted as t₁ and the times the molecule stays in S₂ are denoted as t₂.

The variable time intervals t₁ in Figure 37 are called the *dwell-times* of S₁ and the times intervals t₂ are called the *dwell-times* of S₂. From a single molecule trace we can obtain the average dwell-times <t>, which is the inverse of the sum of rate constants leading away from the state. For the simple example above it is:

$$\langle t_{S1} \rangle = \frac{1}{k_{12}} \quad \text{and} \quad \langle t_{S2} \rangle = \frac{1}{k_{21}}$$

The dwell-time distributions follow an exponential decaying function; short dwell times are more likely than long dwell times. For the simple kinetic scheme underlying Figure 37 the dwell-time distribution of S₁ is given by:

$$f_1(t) = k_{12} \exp(-k_{12} t) = \frac{1}{\tau_{12}} \exp(-t/\tau_{12})$$

And the dwell-time distribution of S₂ is given by:

$$f_2(t) = k_{21} \exp(-k_{21} t) = \frac{1}{\tau_{21}} \exp(-t/\tau_{21})$$

Often time constants τ are used instead of rate constants, with $\tau = 1/k$. These functions are probability density distributions meaning that the probability of finding a dwell-time between t and $t+dt$ is given as $f(t)\,dt$. If more than one reaction leads away from a state, the dwell time distribution for that state becomes a sum of exponential functions. As another example consider a reaction scheme involving three states of which only S_1 is fluorescent (as denoted by the * symbol):

$$S_1^* \underset{k_{21}}{\overset{k_{12}}{\rightleftarrows}} S_2 \underset{k_{32}}{\overset{k_{23}}{\rightleftarrows}} S_3$$

A single molecule trace may look similar to Figure 37, but at the times of low fluorescence we cannot be sure, if the molecule is in state S_2 or S_3. This situation is very common in single molecule kinetics, thus single molecule traces are usually generated by *hidden Markov processes*. The dwell-time distribution of the highly fluorescent state S_1 is the same as previously, while the expected dwell-time distribution of the low fluorescent S_2 state is a sum of two exponentials. Thus analysis of dwell-time distributions from single molecule experiments can give more information about the mechanism than ensemble kinetics.

Any kinetic scheme of unimolecular reactions (leading to first order rate laws) can be described with a matrix approach. This is often used in single molecule kinetics but also in ensemble kinetics. For the example of the reaction scheme above, the system of differential equations is:

$$\frac{dp_{S1}}{dt} = -k_{12}p_{S1} + k_{21}p_{S2}$$

$$\frac{dp_{S2}}{dt} = k_{12}p_{S1} - (k_{21} + k_{23})p_{S2} + k_{32}p_{S3}$$

$$\frac{dp_{S3}}{dt} = \phantom{k_{12}p_{S1} - (k_{21} + k_{23})p_{S2} +\ } k_{23}p_{S2} - k_{32}p_{S3}$$

This can be expressed in vector-matrix notation xxx:

$$\frac{d\mathbf{p}}{dt} = \mathbf{Q}\mathbf{p} \tag{8}$$

(note that vectors are shown in bold lower case and matrices in bold upper case). Equation 8 is a form of the *master* equation in chemical kinetics for first order reactions. It can be used for ensemble kinetics as well as for single molecule kinetics. The matrix of transition probabilities (or rate constants) looks like that:

$$\mathbf{Q} = \begin{pmatrix} -k_{12} & k_{21} & 0 \\ k_{12} & -(k_{21}+k_{23}) & k_{32} \\ 0 & k_{23} & -k_{32} \end{pmatrix}$$

Each element q_{ij} of the matrix $\mathbf{Q}$ specifies the probability per time for a reaction to occur from state i to state j. Each column of $\mathbf{Q}$ must have the sum of zero to fulfil the law of mass conservation. The vector $\mathbf{p}$ contains the time-dependent probabilities of finding each state:

$$\mathbf{p} = \begin{pmatrix} p_{S1} \\ p_{S2} \\ p_{S3} \end{pmatrix}$$

The formal solution of this vector-matrix differential equation is:

$$\mathbf{p}(t) = \mathbf{p}(0)^T \exp(\mathbf{Q}t)$$

Note that $\mathbf{p}(0)^T$ is the transposed form of a column vector, namely a row vector. In the stationary state for t→∞ (infinity), the vector $\mathbf{p}(\infty)$ specifies the probability to find the various states (S_1, S_2, S_3 in our example above). This corresponds to the equilibrium concentrations in ensemble kinetics. While the formal solution seems straightforward, matrix exponentials for 3x3 or larger matrices require a numeric solution.

In case of hidden Markov processes, when there are a number of observable (e.g. fluorescent) and non-observable (e.g. non-fluorescent) states, it is useful to subdivide the matrix $\mathbf{Q}$ into blocks, which only contain transitions between observable/active ($\mathbf{Q}_{AA}$) states and non-observable ($\mathbf{Q}_{NN}$) states:

$$\mathbf{Q} = \begin{pmatrix} \mathbf{Q}_{AA} & \mathbf{Q}_{AN} \\ \mathbf{Q}_{NA} & \mathbf{Q}_{NN} \end{pmatrix}, \text{ which is for the example: } \mathbf{Q} = \begin{pmatrix} -k_{12} & k_{21} & 0 \\ k_{12} & -(k_{21}+k_{23}) & k_{32} \\ 0 & k_{23} & -k_{32} \end{pmatrix}$$

Dependent on the number of states contributing to the block $\mathbf{Q}_{AA}$ and $\mathbf{Q}_{NN}$, the dwell time distribution can be described by just one exponential function, as in the example above for for $\mathbf{Q}_{AA}$ with decay constant $-k_{12}$ (or time constant $\tau_{S1} = -1/k_{12}$) or in general by a sum of exponental functions:

$$f(t) = \sum_{i=1}^{n} a_i \cdot \frac{1}{\tau_i} \cdot \exp(-t/\tau_i)$$

The time constants τ_i can be obtained from the *eigenvalues* λ_i of the n x n sub-matrix $\mathbf{Q}_{AA}$ or $\mathbf{Q}_{NN}$ as $\tau_i = 1/\lambda_i$; the amplitudes a_i are determined by the probability to find a state i in the stationary state:

$$a_i = \frac{p_i(\infty)/\tau_i}{\sum_{k=1}^{n} p_k(\infty)/\tau_k}$$

The equilibrium probabilities can be found from the *eigenvector* corresponding to the eigenvalue of zero (see Box 10 for an explanation of vectors, matrices and eigenvalues).

4 Single-Molecule Kinetics

Box 10: Vectors, matrices and eigenvalues.

Physical quantities such as the location can be specified by numbers such as 5 m and manipulated by multipliers, for example 2 × 5 m. Often specifying a location by a single number is not suffcient; in a three dimensional space we need to give three numbers to specify the location with x-, y- and z-coordinate. These three numbers are conveniently put together into a vector. Vectors are manipulated by multiplication with matrices, for example:

$$\begin{pmatrix} 1 & 2 & 3 \\ 4 & 5 & 6 \\ 7 & 8 & 9 \end{pmatrix} \begin{pmatrix} x \\ y \\ z \end{pmatrix} = \begin{pmatrix} x+2y+3z \\ 4x+5y+6z \\ 7x+8y+9z \end{pmatrix}$$ This results in a new vector.

Vectors can be three-dimensional as in the example above, two-dimensional, one-dimensional (this is a simple number) or in general n-dimensional. The mathematical branch of *linear algebra* deals with n-dimensional vector spaces and their manipulations. There are matrices with certain properties, for example the unity matrix

$$\begin{pmatrix} 1 & 0 \\ 0 & 1 \end{pmatrix} \begin{pmatrix} x \\ y \end{pmatrix} = \begin{pmatrix} x \\ y \end{pmatrix},$$ which does nothing, only multiplication with one.

For certain problems in science it is required to determine the eigenvectors and eigenvalues of a matrix, that are all vectors **x** and eigenvalues λ that fulfill the following equation: **M x** = λ **x**. Eigenvectors are all vectors that give a multiple of themselves, when multiplied by the matrix. For example:

$$\begin{pmatrix} 4 & 1 \\ 3 & 2 \end{pmatrix} \begin{pmatrix} x_1 \\ x_2 \end{pmatrix} = \lambda \begin{pmatrix} x_1 \\ x_2 \end{pmatrix}$$

There are two resulting eigenvalues and eigenvectors for this example:

For λ = 1, $\mathbf{x} = \begin{pmatrix} 1 \\ -3 \end{pmatrix}$ and for λ = 5, $\mathbf{x} = \begin{pmatrix} 1 \\ 1 \end{pmatrix}$

Note that multiples of the above eigenvectors would be valid solutions, such as $\begin{pmatrix} 2 \\ -6 \end{pmatrix}$ for λ = 1. The are various computer software (Octave, Mathematica™, Matlab™) that can carry out this and many more linear algebra calculations.

4.1.2 Bimolecular reactions

Consider the following bimolecular elementary reaction:

$$E + S \xrightarrow{k_{12}} ES$$

The ensemble reaction rate is given as the second order rate law $d[ES]/dt = k_{12}[E][S]$ with the rate constant in the units of L mol^{-1}s^{-1}. The physical basis for the units of the rate constant, volume/(mol time), is the volume swept out by the two potentially colliding molecules when they move through space; $k_{12}\, dt$ in the units volume/mol is proportional to this collision volume.

The stochastic rate equation is expressed with the number of molecules n_E, n_S:

$$\frac{dn_{ES}}{dt} = \alpha_{12} n_E n_S \quad \text{or expressed with probabilities} \quad \frac{dp_{ES}}{dt} = \alpha_{12} p_E p_S$$

The stochastic rate constant α_{12} includes the probability per time that a collision between two molecules occurs, i.e. $\alpha_{12}\, dt$ is the probability that a collision occurs in the time interval dt (and taking into account that not every collision leads to a reaction) ; it does not include the volume. This collision probability is inversely proportional to the system volume V, i.e. the larger the volume the two molecules are inside the less likely a collision is. Thus the stochastic rate constant for a bimolecular reaction in the system volume V is given as:

$$\alpha_{12} = \frac{k_{12}}{V N_A}$$

(in the units of molecules/time with the Avogadro constant $N_A = 6.0225 \cdot 10^{23}$). For the trimolecular reaction, although they hardly occur in practice, the relationship would be $\alpha = k/(V^2 N_A)$. As a consequence for bimolecular reactions in small volumes such as a cell, the stochastic rate constants are quite different from the ensemble rate constants.

Another peculiarity occurs with bimolecular association reactions of the type:

$$A + A \xrightarrow{k_{12}} A_2$$

The stochastic rate equation is:

$$\frac{dn_{A2}}{dt} = \frac{1}{2} \alpha_{12} n_A (n_A - 1) \quad ; \text{with } n_A = \text{number of A molecules in the system volume}$$

This is based on counting the number of distinct pairs of 'A' molecules. For example in case of two molecules of A, we only have one pair that can collide, while in case of three molecules we have three pairs (1-2, 1-3 and 2-3). Thus the stochastic rate constant in this case is: $\alpha_{12} = \dfrac{2k_{12}}{VN_A}$ (in the units of molecules/time)

The kinetics of a system of biomolecular reactions can be treated formally with a master-equation approach. However, this approach does not does not lead to the vector-matrix equation as shown in equation 8. Therefore, bimolecular systems are usually treated with *pseudo-first order* kinetics (see 2.2.2), in which the concentration of one the reactants is considered as practically constant. The Michaelis-Menten reaction scheme can be written in pseudo-first order approximation:

$$E \underset{k_{-1}}{\overset{k_1[S]}{\rightleftarrows}} ES \xrightarrow{k_2} P + E$$

The matrix of transition probabilities is:

$$\mathbf{Q} = \begin{pmatrix} -k_1[S] & (k_{-1}+k_2) & 0 \\ k_1[S] & -(k_{-1}+k_2) & 0 \\ 0 & k_2 & 0 \end{pmatrix}$$

The conservation of mass does not apply to this reaction scheme, as the substrate S does apparently not participate in the reaction, it becomes part of an apparent first order rate constant $k_1' = k_1[S]$. At the same time product appears, which is not balanced by the consumption of substrate, as [S] is assumed to be constant.

4.2 Single molecule enzyme kinetics

Contrary to the initial rate approach in ensemble-based enzyme kinetics, in single molecule kinetics the probability density functions of dwell-times can be obtained by measuring many single molecule traces. This is equivalent to measuring the time-dependent concentration of intermediates in ensemble kinetics, thus it provides much more detail compared to initial rate measurements. As an example we consider the oxidation of the substrate cholesterol (S) to cholestenone (P, cholest-4-en-3-one) catalysed by cholesterol oxidase:

$S + O_2 \rightarrow P + H_2O_2$

Cholesterol oxidase contains the cofactor FAD, which is fluorescent in its oxidised (FAD) state and non-fluorescent it its reduced (FADH₂) state. This reaction was observed with an inverted laser-scanning fluorescence microscope and a typical observation is shown in Figure 38.

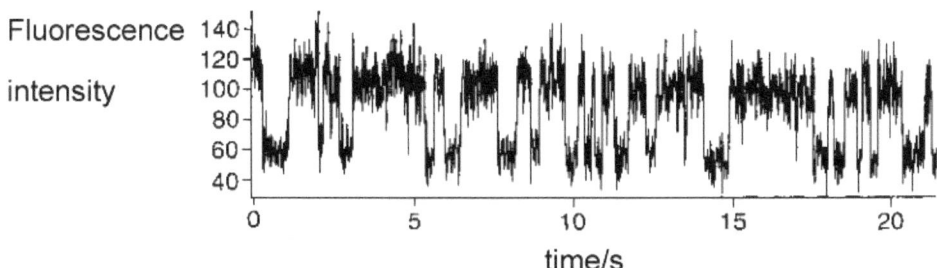

Figure 38: Extract of a real time observation of the enzymatic activity of agarose-gel immobilised cholesterol oxidase monitored by fluorescence microscopy at 0.2 mM cholesterol and 0.25 mM oxygen [from (Lu, Xun & Xie, 1998). Reprinted with permission from AAAS]

The reaction mechanism for the cholesterol oxidation catalysed by cholesterol-oxidase is as follows:

$$\text{E-FAD*} \underset{k_{-1}}{\overset{k_1[S]}{\rightleftharpoons}} \text{E-FAD·S*} \overset{k_2}{\rightarrow} \text{E-FADH}_2 + \text{P}$$

$$\text{E-FADH}_2 \underset{k'_{-1}}{\overset{k'_1[O_2]}{\rightleftharpoons}} \text{E-FADH}_2\text{·O}_2 \overset{k'_2}{\rightarrow} \text{E-FAD*} + \text{H}_2\text{O}_2$$

The star (*) symbol denotes the highly fluorescent state that is observed in Figure 38, which is a combination of two states. Thus the distribution of 'on-times' $p_{on}(t)$ ($p_{on}(t)$ dt is the probability to find an on-time between t and t+dt) is given by the convolution two exponentials, for example with $k_1 \cdot 0.2$ mM = 2.9 s^{-1} and k_2 = 17 s^{-1} (k_{-1} was assumed to be zero), which corresponds to E-FAD* and E-FAD·S*. A similar analysis was performed for the off-times (Lu, Xun & Xie, 1998). Interestingly, among 30 different single enzyme molecules the authors found a variation of k_2 between 3 s^{-1} and 14 s^{-1} for a cholesterol derivative as substrate. This *static disorder* in the kinetic properties was attributed to small chemical modifications of the enzymes studied caused for example by proteolysis or oxidation. For one single enzyme molecule studied, the authors examined the correlation between on-times separated by a number of turnover cycles. Normally there should be no correlation, but correlation was found between on-times separated by up 10 turnover cycles and attributed to heterogeneity of rate constant k_2. While this *dynamic disorder* could be considered as an invalidation of the Markov-process assumption, it can be accounted for within the Markov model by considering additional conformational states of the enzyme. It is conceivable that a distribution of different long-lived conformational states exist with different rate constants. However, other experiments with different enzymes have not confirmed dynamic disorder as a general principle, but suggested instead that the low signal/noise ratio in single molecule experiments may lead to artifacts (Terentyeva et al, 2012).

4.2.1 The single molecule Michaelis-Menten equation

From the pseudo-first order Michaelis-Menten scheme shown in chapter 4.2.1 it is possible to derive an expression analogue to the initial reaction rate of the Michaelis-Menten equation. In the single molecule case an equivalent to the initial reaction rate is the inverse of the average waiting time <t> for the formation of a product molecule. This average is found by integrating over the product between t and probability density function of the dwell-times $f(t)$ for product formation:

$$\langle t \rangle = \int_0^\infty t f(t) dt$$

Using this approach it has been shown that (Kou et al, 2005):

$$\frac{1}{\langle t \rangle} = \frac{k_2 [S]}{[S] + K_M} \qquad (9)$$

This is the single molecule Michaelis-Menten equation. As explained in more detail in chapter 4.3 algorithms for stochastic simulation of single molecule experiments have been developed. An example for such a simulation of the Michaelis-Menten scheme with one enzyme and 100 substrate molecules is shown in Figure 39.

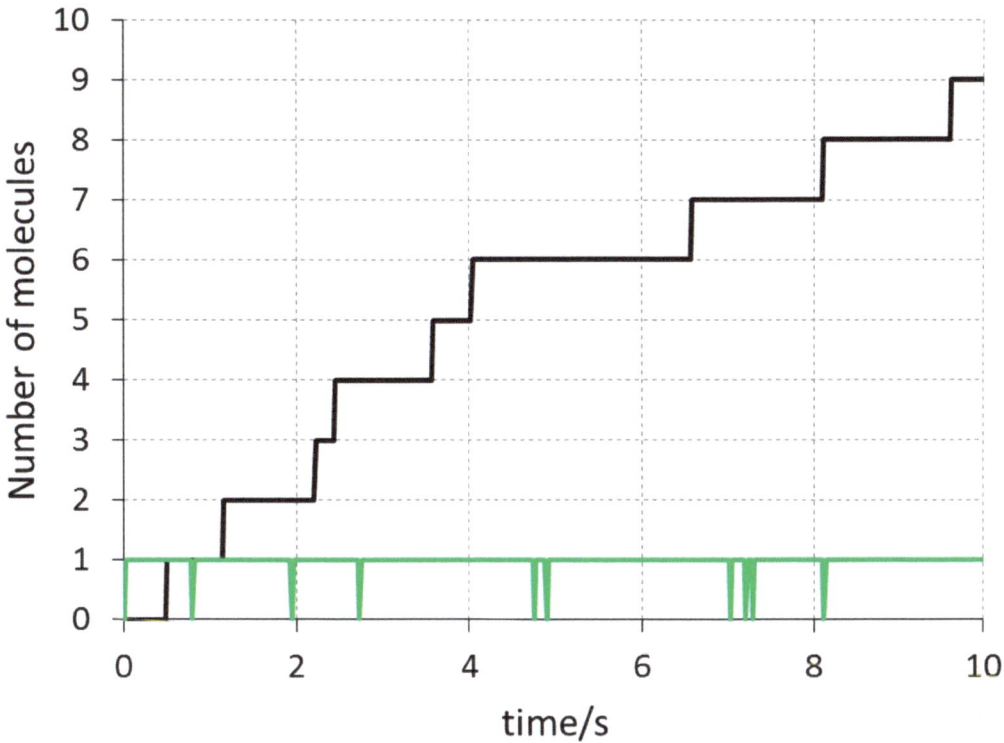

Figure 39: Results of a stochastic simulation of Michaelis-Menten mechanisms showing the number of product molecules (black curve) and the number of ES complexes (green curve). Note that due an excess of substrate we have steady state conditions. The simulation was conducted with k_1 = 10 molecules/s, k_{-1} = 10 s^{-1}, k_2 = 1 s^{-1}. The starting conditions were 1 enzyme molecule and 100 substrate molecules in a volume of 1 fL. The ensemble value for the rate constant k_1 is then 6·10^9 L mol^{-1} s^{-1}. The simulation was performed with COPASI (Hoops et al, 2006)

For the simulation conditions in Figure 39, the Michaelis constant K_M is

$$K_M = \frac{k_{-1} + k_2}{k_1} = \frac{10\,\text{s}^{-1} + 1\,\text{s}^{-1}}{10\,\text{molecules} \cdot \text{s}^{-1}} = 1.1 \text{ molecules}$$

(in a volume of 1 fL). Thus at a number of 100 substrate molecules the expected average waiting time can be obtained from equation 9 as $\langle t \rangle$ = 1.01 s. Analysis of the (short) simulation trajectory yields an average waiting time of 1.07 s.

4.3 Stochastic simulations

While numerical integration of differential equations in ensemble kinetics is sometimes referred to erroneously as simulation, the stochastic behaviour of single molecules cannot be obtained from numerical integration. Stochastic simulations take into account each molecule in the system and the various reactions (or transitions) available to each molecule. The original stochastic simulation algorithm for chemical kinetics was suggested by Gillespie (1977). Consider a molecule A that can undergo various reactions (described by rate constants $k_1, k_2, .., k_i$):

$$A \xrightarrow{k_1} \quad A \xrightarrow{k_2} \quad A \xrightarrow{k_i}$$

The Gillespie algorithm entails drawing a random number that is exponentially distributed with the decay constant $\sum k_i$; this specifies the time the molecule stays in state A, which is on average $<t> = 1/\sum k_i$. Another random number determines, which reaction takes place, while the probability for each reaction r is given as $k_r/\sum k_i$. The algorithm is summarised in Box 11. The original algorithm by Gillespie, often called the direct method, has been modified to increase the speed of the simulation as well as approximate methods with further performance increase, such as tau-leaping methods, have been developed (Gillespie, 2007).

Box 11: The Gillespie algorithm for stochastic simulations.

Initialise (set initial numbers)

⮡ Draw random variable t (exponentially distributed with $<t> = 1/\sum k_i$)*
   Draw another random integer variable to determine which reaction occurs
       (the probability for each reaction r is: $k_r/\sum k_i$)*
   Update molecule numbers
   Increment time $T = T + t$
⮡ Until specified end-time is reached.

*) Note that in case of second order reactions the instantaneous number of molecules of one of the reaction partners would be taken into account, e.g. $k_i\ n_A\ n_B$ for the reaction A + B → …

Key Points: Chapter 4 Single molecule Kinetics

- Single molecule unimolecular rate constants are the same as their ensemble counterparts.
- Single molecule bimolecular rate constants are inversely proportional to the system volume.
- Vector-matrix equations are used to summarise systems of first-order (or pseudo-first order) rate equations.
- Single molecule enzyme kinetics shows heterogeneity of rate constants.
- The single-molecule Michaelis-Menten equation is given as

$$\frac{1}{\langle t \rangle} = \frac{k_2[S]}{[S]+K_M}$$

- The Gillespie-algorithm can be used for simulation of single molecule traces for arbitrary reaction schemes.

Reading for chapter 4:

Gillespie DT (1977) Exact Stochastic Simulation of Coupled Chemical-Reactions, *J Phys Chem-Us* **81:** 2340-2361

Gillespie DT (2007) Stochastic simulation of chemical kinetics, *Annu Rev Phys Chem* **58:** 35-55

Hoops S, Sahle S, Gauges R et al (2006) COPASI: A COmplex PAthway SImulator, *Bioinformatics* **22:** 3067-3074

Kou SC, Cherayil BJ, Min W et al (2005) Single-molecule Michaelis-Menten equations, *Journal of Physical Chemistry B* **109:** 19068-19081

Lu HP, Xun LY, Xie XS (1998) Single-molecule enzymatic dynamics, *Science* **282:** 1877-1882

Makarov DE (2015) *Single Molecule Science: Physical Principles and Models*, 1st edn. London: CRC Press (Taylor and Francis).

Terentyeva TG, Engelkamp H, Rowan AE et al (2012) Dynamic Disorder in Single-Enzyme Experiments: Facts and Artifacts, *Acs Nano* **6:** 346-354

ABOUT THE AUTHOR

After his PhD in biophysical chemistry at the University of Bielefeld, Germany, the author Andreas Kukol conducted postdoctoral research for three years in the Department of Biochemistry, University of Cambridge, UK. Then he joined the Department of Biological Sciences, University of Warwick, as lecturer and moved in 2007 to the School of Life and Medical Sciences, University of Hertfordshire, UK. His teaching covers biochemistry at undergraduate and postgraduate level, while his research uses methods of computational biochemistry and bioinformatics applied to antiviral drug discovery, transmembrane protein-protein interactions that control cardiac function and protein-protein interactions in plant-pathogen systems.

www.ingramcontent.com/pod-product-compliance
Lightning Source LLC
Chambersburg PA
CBHW041314180526
45172CB00004B/1089